NUCLEAR POWER: BOTH SIDES

Nuclear Power: Both Sides

The Best Arguments For and Against
the Most Controversial Technology

EDITED BY MICHIO KAKU, PH.D.
AND JENNIFER TRAINER

W • W • Norton & Company

NEW YORK • LONDON

Copyright © 1982 by Michio Kaku and Jennifer Trainer
All rights reserved.

Printed in the United States of America.

The text of this book is composed in V-I-P Times Roman, with display type set in Photo-typositor Schadow Werk.
Manufacturing by the Maple-Vail Book Manufacturing Group.
Book design by Winston Potter

First published as a Norton paperback 1983

Library of Congress Cataloging in Publication Data

Main entry under title:

Nuclear power, both sides.

    Includes bibliographical references and index.
    1. Atomic power—United States.   I. Kaku,
Michio.   II. Trainer, Jennifer.
TK9023.N8   333.79′24′0973   82–6426

ISBN 0-393-01631-5 AACR2

ISBN 0-393-30128-1 pbk

W. W. Norton & Company, Inc. 500 Fifth Avenue, New York, NY 10110
W. W. Norton & Company Ltd. 37 Great Russell Street, London WC1B 3NU

6 7 8 9 0

Contents

Introduction   7

Acknowledgments   9

1. A Brief History

The Early Years   *Boyd Norton*   15

2. Radiation

Introduction   27

Underestimating the Risks   *Karl Z. Morgan*   35

Protecting the Public   *Allen Brodsky*   46

George Orwell Understated the Case

    John W. Gofman   56

Exaggerating the Risks   *Bernard L. Cohen*   69

3. Reactor Safety

Introduction   81

Safe Enough   *Anthony V. Nero, Jr.*   86

Second Thoughts   *Jan Beyea*   97

4. Nuclear Waste Disposal

Introduction   109

No Technical Barriers   *Fred A. Donath*   115

Will It Stay Put?   *Robert O. Pohl*   123

5. Economics

Introduction   135

The Industry's Worst Enemy   *Ralph Nader*

    and Richard P. Pollock   141

On the Road to Recovery   *Tony Velocci, Jr.* 148
Living Without Nuclear Energy   *Vince Taylor* 155
The Effects of a Nuclear Phase-out
            Alan S. Manne and Richard G. Richels 160

6. Beyond Light-Water Reactors

Introduction   171
Soft Energy Paths   *Amory B. Lovins*
                        and L. Hunter Lovins 176
A Golden Decade for Solar Energy?
                    Modesto A. Maidique 192
Closing in on Fusion   *Stephen O. Dean* 198
Breeder Reactors: The Next Generation
            Robert Avery and Hans A. Bethe 206

7. Where Do We Go From Here?

Introduction   217
The Future of Nuclear Energy
                    Alvin M. Weinberg 219
Denuclearization   *Richard Falk* 226
The Antinuclear Movement   *David Dellinger* 233
The Hidden Agenda   *Bertram Wolfe* 238

Epilogue   247
Notes   251
Index   269

Introduction

Are you pro- or antinuclear?

Nearly everyone has been asked that question, and whether a person advocates or denounces the commercial use of nuclear power is likely to engender intense debate. Indeed the issue has become so polarized and emotionally charged that the question is often posed as though one's political and philosophical ideologies might be determined by the response. Sometimes they can be.

Although many scientific experts have established themselves on one side or the other of the nuclear fence, there remains a large group of people who are undecided about the technology. Some no longer trust the experts and are confused by conflicting scientific reports. Others want to believe that nuclear power is safe, but in an age when technological innocence was lost long ago, people are more critical of what is being done in the name of progress. With regard to nuclear power, they need the facts and opinions explained to them in a clear, logical manner.

Months ago we decided to create a book on nuclear power that might serve as a guide for the perplexed citizen. We believed that only a book providing the most cogent arguments from *both* sides could clarify the complexity of the issue. Our goal was to present the facts as accurately and comprehensively as possible so that people might draw their own conclusions about the consequences of living with—or without—nuclear power.

In order to best represent both sides, we asked a number of extraordinary authorities to contribute essays to the book. In these decisive, sometimes blistering, essays a wide array of physicists, economists, industrialists, environmentalists, geologists, university professors, and laboratory heads present valid persuasive arguments for and against nuclear power. Not only do they lay out their opinions, substantiating each point with scientific evidence, but they address their opponents' claims as well.

Our job as editors was to integrate the material, provide introductions to each chapter, and cut through the jargon jungle of economic and scientific terms that often make books on the subject impenetrable to the layman. Our combined experience—as a nuclear physicist and as a professional writer—worked well in this regard.

Locating these specialists was our first task. Next we had to convince them that a book which addressed one of the most volatile, convoluted issues of the decade could possibly be balanced. Some were delighted at the opportunity to contribute to a book that would give them a valid forum by which to reach the general public. Others were initially skeptical about Michio Kaku's affiliation with the antinuclear movement. We therefore attended nuclear conventions, spoke at meetings and forums, and visited nuclear physics laboratories to convince potential contributors of our objectivity and devotion to the same elusive quality that we demanded of them: intellectual honesty.

To further guarantee that the book would be scrupulously fair and scientifically valid, we established a series of checks and balances. For example, both pro- and antinuclear essays were read and appraised by several other contributors adhering to the same point of view. Our introduction to each chapter was approved by all whose essays appeared in the chapter. Other specialists who were not contributing directly to the book—scientists, energy analysts, working associates—critiqued the introductions and essays as well. The essays were not, however, exchanged between pro- and antinuclear contributors; had we done this the book might have become mired by endless rebuttals and counterrebuttals.

To elicit the most evocative response from each contributor, we posed questions that concentrated on—and often challenged—that person's professional views. We asked consumer advocate Ralph Nader, for example, about the costs of nuclear plants to consumers. "You favor more regulation of private corporations," we said, "and yet the nuclear regulatory process has become so politicized that power plants are shut down for symbolic reasons, rather than for actual safety hazards, at a cost of $800,000 per day. In the end, won't the consumer pay the most for the heavy hand of regulation?" To Allen Brodsky, a health physicist for the U.S. Nuclear Regulatory Commission, we wrote: "It is not enough to say that the experts conclude that the general public is in no danger from routine emissions from nuclear power plants. People have heard it before and no longer believe it. Instead, how can you be sure that current exposure limits afford adequate protection to the public? What kinds of studies have been performed?"

Although the contributors differed strongly about many scientific matters, they all agreed on one point: that this book has long been needed. Without remuneration they devoted many hours to the preparation of their essays, often revising and graciously accepting editorial suggestion, in a true spirit of cooperation toward a mutual endeavor.

The reader should not expect one side to "win" the argument, for the book is not intended to be a debate. Rather it is a source book of information to illuminate the many technical, political, and economic aspects of nuclear power that make the controversy so complex. Basically, people want to know how nuclear power affects their lives, and we have tried to tell them. Some antinuclear activists may think that the book is pro-nuclear, and among nuclear advocates there will be those who find the book antinuclear. If that proves to be the case, we will have done our work well.

Jennifer Trainer
Michio Kaku

New York
January 1982

Acknowledgments

First we want to thank our contributors, who devoted an enormous amount of time and effort to this project. We also greatly appreciate the assistance and thoughtful advice of numerous professionals in and out of the energy field, in particular Irvin C. Bupp of the Harvard Graduate School of Business Administration, Joel Darmstadter of Resources for the Future, Charles Komanoff of Komanoff Energy Associates, and Isaac Asimov. Finally, we offer our special thanks to Carol Houck Smith, Elizabeth De Cresenzo, Jeanie Chin, and Bradley Olsen for their unwavering support.

NUCLEAR POWER: BOTH SIDES

Commercial Nuclear Power in the United States

● 72 reactors licensed to operate

▲ 89 reactors being built

SCALE IN MILES

0 25 50   100   200   300

1

A Brief History

The Early Years
BOYD NORTON

The discovery of nuclear energy has been hailed as one of our greatest scientific triumphs and assailed as the ultimate Faustian bargain. The following brief history of nuclear power calls to mind the progression of events—some etched permanently in memory, others long forgotten—that brought nuclear power into being. Explaining the history is a nuclear physicist who is also an accomplished writer. Boyd Norton began his career in 1960 as a reactor physicist with the Atomic Energy Commission's National Reactor Testing Station in Idaho Falls, Idaho, where he performed studies on early test reactors. In 1969 he left the field of nuclear physics and became a freelance photographer and writer. He is the author of six books, including *Wilderness Photography* and *Alaska: Wilderness Frontier,* as well as numerous magazine articles.

Atomic energy and I grew up together.

I remember well the headlines and stories. For an impressionable nine-year-old, already a whiz kid in science and determined to be a chemist or physicist, it was all very exciting: an atomic bomb, deriving its enormous energy from a mysterious new process called nuclear fission. Adding to the fascination and mystique was the fact that the bomb had been developed as part of a super-secret project and revealed to the world for the first time with the deadly detonations over Hiroshima and Nagasaki. In 1945 it all seemed so clear and simple: nuclear fission was to become the energy source of the future. The genius that had gone into creating the atomic bomb could clearly find ways to utilize nuclear power safely for peacetime uses.

The process of nuclear fission is awesome, beautiful, elemental, and elegant; it gets down to the very roots of the universe. The discovery, exploration, and utilization of nuclear fission is one of mankind's greatest intellectual achievements.

It began in the 1930s, an exciting era for the world of physics. In 1932 Sir James Chadwick discovered the neutron, an ethereal subatomic particle that provided physicists with a means of probing still farther into the mysteries of the atomic nucleus.*

---

*The atom is comprised of a tiny nucleus surrounded by one or more shells of electrons. While the atom is so infinitesimal that 100 million atoms lined up would measure only one centimeter, the nucleus is 100,000 times smaller than the atom itself. Inside the nucleus are two types of particles occurring in approximately equal numbers: the protons, which carry the positive electric charge of the nucleus, and the neutrons, which weigh approximately the same as protons but carry no charge.

15

In 1939 two Austrian physicists, Otto Hahn and Fritz Strassman, demonstrated that the nucleus of uranium atoms actually split, or fissioned, into two lighter parts when hit by neutrons. Normally, because all positive charges repel each other, the electrical barriers between nuclei are sufficient to prevent them for colliding with each other. But because the neutron carries no charge it can easily penetrate the electrical barrier of the nucleus and splinter the nucleus. More important was the discovery that even a small quantity of uranium could release, through fission, thousands of times more energy than could be derived from combustion or other chemical reactions.* Almost immediately the military potential of this new process was recognized.

The discovery of nuclear fission took place as World War II was imminent. Much of the initial research was carried out in Nazi Germany, though in time many of the European scientists fled to settle in the United States, the United Kingdom, and Canada. One of those was Leo Szilard, an energetic Hungarian physicist who understood thoroughly the military potential of a fission bomb. Fearing the consequences if Nazi Germany should develop such a weapon (Hilter's war machine had already sealed off the uranium mines of Czechoslovakia), Szilard urged his friend, Albert Einstein, to send a letter to President Roosevelt outlining the potential for the development of a fission bomb. Although Einstein would later regret it, his letter played a pivotal role in moving the Roosevelt administration to embark on a massive government project to build the atomic bomb. Hundreds of the world's top scientists were assembled in top secrecy in perhaps the greatest scientific and engineering undertaking of all time: the Manhattan Project.

The events of the Manhattan Project can best be described as a curious blend of luck, inspired direction, dedicated research, brilliant scientific deduction, and more luck. The obstacles—physical, scientific, and economic—were enormous. Using untried methods on unseen substances with uncertain properties, unknown materials were to be refined by unspecified processes to produce untested components for an unimaginable weapon. Failure was unacceptable.

Considering the timetable of events, the accomplishments of the Manhattan Project were doubly miraculous. Only three years after the discovery of fission the first nuclear reactor was built and tested successfully in a converted squash court at the University of Chicago. A brilliant Italian physicist named Enrico Fermi, whose suggestion in 1934 had led to the discovery of fission, directed this historic experiment in 1942. Three years after the first reactor was operated, an atomic bomb was successfully detonated in the lonely New Mexico desert. In many ways it was like making a successful lunar landing only six years after the invention of the first crude rocket (see "Building the A-Bomb").

---

* Uranium was the key material because it fissions easily and releases two to three more neutrons in the process. These neutrons, in turn, split more uranium atoms, which can build into a chain reaction so rapidly that within a fraction of a second, trillions of atoms are undergoing fission.

BUILDING THE A-BOMB

According to Einstein's formula $E = mc^2$, even a small amount of mass (m) inside the atom can be magnified by a huge number (c^2, or the speed of light squared) to create enormous amounts of energy (E). Theoretically this energy can be released whenever the uranium atom is split. In practice, however, releasing the energy in a bomb was not easy. There were two major challenges to be surmounted.

First, two types of uranium atoms exist, which are chemically identical and difficult to separate. Almost all naturally occurring uranium is U-238 (with 92 protons and 146 neutrons in the nucleus), which does not fission easily and cannot sustain a chain reaction. Its exotic cousin U-235 (with three fewer neutrons) fissions readily and is the active ingredient in the atomic bomb; U-235 is also quite rare, comprising only 0.7 percent of all naturally occurring uranium. In order to develop a bomb, scientists had to increase, or *enrich,* the concentration of the exotic fissionable U-235 and reduce the concentration of the more plentiful U-238. Any amount less than 20 percent enriched U-235 would be too low for the purpose of making an atomic bomb.

The second problem was to manufacture enough U-235—known as a critical mass— for the bomb to ignite. Any amount less than critical mass (approximately 25 pounds) would cause too many neutrons to leak out of the uranium. If, however, enough of the U-235 could be held together for a split second, the temperature would have time to rise to several million degrees Centigrade, at which point the uranium would vaporize and the gases develop enormous pressures. In a fraction of a second the gases would expand, creating a tremendous explosion: the result, a nuclear bomb.

The design of the atomic bomb was surprisingly simple. First two pieces of uranium-235, each less than critical mass, were separated. Then one piece, in the form of a missile or bullet, was propelled toward the other rapidly by high explosives. When the two pieces were slammed together, a critical mass was formed, which instantly initiated the chain reaction and the atomic explosion.

Another type of bomb was made from plutonium, a fissionable man-made element. Because of certain technical complications, however, the critical mass for the plutonium had to be assembled much faster than for U-235. Thus the "gun-type method" used for U-235 could not be used for plutonium. Instead, the subcritical plutonium core, in the form of a sphere, was surrounded by TNT which, when detonated, caused a high-pressure shock wave, which squeezed the plutonium into a dense, supercritical mass. In less than a millionth of a second the chain reaction began and caused an atomic explosion. This is called the "implosion method." The first atomic bomb tested in New Mexico in 1945 was of this implosion type, using plutonium, while the Hiroshima bomb was of the "gun type," using U-235.

With the end of World War II many scientists who had worked diligently on the atomic bomb, along with officials of American corporations who had served as contractors to the Manhattan Project, expected that the money and manpower used for developing the bomb would be redirected toward peacetime uses of fission. But right from the start it seemed nearly as difficult to separate the military atom from the peaceful atom as it had been to separate the fissionable isotope uranium-235 from the nonfissionable U-238.

In 1946 Congress passed the Atomic Energy Act, which established the Atomic Energy Commission (AEC) to develop peaceful atomic programs and the Congressional Joint Committee on Atomic Energy (JCAE) to oversee the AEC's activities. The act also attempted to prohibit any exchange of nuclear technology, even between the United States and her wartime Allies, until international safeguards against nuclear weapons could be established. For a short while after World War II the U.S. had a monopoly on atomic weaponry. But in 1949 the Soviet Union exploded its first nuclear device and the nuclear arms race was on. The AEC gave priority to military research, with only token funding of programs useful to developing nuclear power as a commercial, peacetime source of energy.

The 1950s: The Peaceful Atom

In the early 1950s the new Eisenhower administration revised the nation's energy policy, which had effectively prohibited the private sector from owning and operating nuclear reactors. In December 1953 Eisenhower announced the long-awaited Atoms for Peace program, which called for agreements with other countries to share technical and scientific expertise.

Domestically, however, several major obstacles hindered the transfer of the peaceful atom to the private sector. First, much of the nuclear technology remained highly classified, and companies not involved directly in military research had restricted access to important information. Second, vast engineering problems had not yet been solved; it was extremely difficult to create an entirely new technology of steel and rare metals which could withstand the blistering conditions of intense radiation, temperature, and pressures found in reactors. Third, the utilities were not convinced that reactors could generate electricity competitively with proven technologies like coal and oil. For the utilities, one crucial question remained: Could the atom generate profits?

Meanwhile by the early 1950s the AEC was being victimized by political philosophies. Lewis Strauss, chairman of the AEC through much of the Eisenhower administration, was a staunch conservative who believed that the responsibility for commercial development of nuclear power lay with private industry, not the government. He was often opposed by New Deal Democrats in the JCAE who, among other

things, wanted the federal government to subsidize the construction of full-scale demonstration reactors for every major reactor type.

Whereas the JCAE Democrats wanted the government to fully control the development and implementation of a peaceful nuclear program, Strauss wanted the government to underwrite the research and development costs, then hand the technology
over to private industry. Despite these conflicts of interest several incentives were
offered to the industry during those years. For example, in 1952 the government
funded the National Reactor Testing Station in Idaho to develop materials that could
withstand the intense radiation of a nuclear reaction. In 1954 Congress revised the
Atomic Energy Act to authorize private ownership of nuclear facilities, under AEC
licenses. In 1957 Congress passed the Price-Anderson Act, which served to protect
utility companies from full financial liability for a serious nuclear accident, should
one occur. The act set a ceiling of $560 million recoverable in damages, with the
utilities assuming $60 million of the liability and the federal government the remain-
ing $500 million. (In 1975 the act was amended to virtually eliminate the govern-
ment's role in this no-fault insurance plan.*)

But it also took the pioneering efforts of Admiral Hyman G. Rickover to induce
industry executives to consider seriously the prospects of commercial nuclear power.
A navy man, Rickover used his considerable influence in both the AEC and the
navy's Bureau of Ships to cut through the bureaucratic red tape and launch the *Nautilus,* the world's first nuclear submarine, in 1954. Utility executives and manufacturers were dazzled by Rickover's light-water design;† while the British built the
clumsy, inelegant carbon dioxide–cooled reactor, and the Canadians, hampered by
their lack of uranium enrichment plants, settled for the heavy-water reactor, Rickover's submarines were cruising effortlessly under the North Pole.

Buoyed by the success of the nuclear submarine, Rickover took the remnants of a
project for a nuclear-powered aircraft carrier and organized a program which eventually culminated in the construction of the world's first full-scale nuclear generating
plant in 1957 at Shippingport, Pennsylvania. Designed by Westinghouse and paid for
by the federal government, this 60-megawatt demonstration light-water reactor was
essentially a scaled-up version of the small submarine reactor. When the first nuclear-
generated electricty was sent coursing through the power grid, commercial nuclear
power had finally been weaned from the military program.

---

*The amended version of the act also sets a $560-million ceiling on liability for a nuclear accident,
whereby $160 million is provided by private insurers, $5 million per reactor is provided by the utilities,
and the rest is insured by the federal government. With 72 licensed operating reactors in 1982, the federal
government's share is only $40 million ($[560 - 160] - [5 \times 72] = 40$). Of course in the event of an accident
Congress can always appropriate funds beyond the $560-million limit.
† The light-water reactor is just one of many reactor types. The core is cooled by ordinary water (H_2O).
By contrast, in a heavy-water reactor or a gas-cooled reactor, the core is cooled by heavy water (D_2O) or
a gas (such as CO_2, carbon dioxide), respectively.

The 1960s: The Great Bandwagon

The early 1960s marked the beginning of what some historians have called the "Great Bandwagon Market" for nuclear power. Essentially, a gold rush to produce nuclear electricity began. The operation of the Shippingport reactor was cited as proof that safe, economical commercial nuclear power was at hand. In 1963 Jersey Central Power and Light Company contracted for the purchase of a 515-megawatt reactor to be built at Oyster Creek, New Jersey. It would be the first plant to generate electricity without benefit of federal subsidy. In other words, JCPL's economic analysis had indicated that nuclear fission was the cheapest source of power for that region. In addition the contractor, General Electric, had offered the plant at a fixed cost, which would change only in accordance with inflation over the construction period. Soon the other three American reactor manufacturers (Babcock and Wilcox, Westinghouse, and Combustion Engineering) were offering reactors at a firm, guaranteed price, thus providing a great incentive for utilities to go nuclear. Because the manufacturers would deliver complete generating stations to the owner, which merely had to "turn the key" to start the generating equipment, these early projects were called "turnkey" reactors.

By 1965 the Great Bandwagon Market for nuclear power was in full swing. Between 1965 and 1967 utilities placed orders for fifty plants totaling 40,000 megawatts of electrical generating capacity. The four reactor manufacturers were tripping over each other to offer utility companies the lowest bid. But while the decade was a time of incredible success for the nuclear power industry, it also saw the culmination of a ten-year struggle between the AEC and JCAE.

The JCAE Democrats were frustrated that Lewis Strauss wanted industry, not government, to take primary responsibility for peaceful atomic development. In 1964 they were vindicated when Milton Shaw was appointed director of the AEC's Division of Reactor Development and Technology. A protégé of Admiral Rickover, Shaw was an aggressive, no-nonsense administrator who believed that the government should vigorously develop a strictly controlled nuclear program. High on Shaw's priority list was the experimental fast breeder reactor.* Shaw had the foresight to see that the light-water system wasted uranium, which was in limited supply, and that the breeder reactor could stretch fuel supplies indefinitely.

Like many officials in his day, Shaw thought that by 1964 the presence of light-water reactors on the market was proof that the safety and design of the system was complete. He had little interest in further safety research on light-water reactors, and diverted funds and manpower toward the breeder. However, the breeder program ran into some difficulties. A consortium of utilities, led by Detroit Edison, began the

---

*The breeder reactor creates more fuel than it consumes by turning "waste" U-238 into valuable plutonium-239. The breeder reactor consists of a 10–20 percent enriched core of uranium and plutonium surrounded by a shell, or "blanket," of U-238. The intense neutron radiation emitted by the core is absorbed by the U-238 in the blanket, which then transmutes into plutonium-239.

ambitious project of building the first commercial breeder reactor—Fermi I—in the early 1960s. On October 5, 1966, a few days after it began to generate electricity, the Fermi reactor suffered a small melting incident in its uranium core due to a jamming in its sodium coolant system. Although only 2 percent of the core melted, and the reactor was probably in no danger of a meltdown, the accident did exceed the maximum credible accident postulated by the industry. Repair operations took more than a year. Four years later the Fermi I reactor was about to go into operation again when a small sodium explosion took place in the reactor pipes. Eventually, enormous delays, escalating costs, and unexpected technical difficulties caused the AEC to lift Fermi I's license in 1972. To this day no commercial breeder reactor is operating in the United States.

Meanwhile the light-water reactor design pioneered by Rickover was being scaled up, from the 60-megawatt *Nautilus* submarine and Shippingport reactor to the 600-, 900-, and finally 1000-megawatt reactors of today. (See "How Reactors Work.") Had Shaw been more sympathetic toward safety research as the light-water reactor was being scaled up five, ten, twenty times, the gap in our understanding of reactor accidents might have been closed. "One of the grievous errors of the Atomic Energy Commission in the 1960s," said Dr. Carroll Wilson, the first general manager of the AEC during the 1950s, "was the failure to carry out the repeatedly recommended experiment of running a [large-scale] reactor to destruction to find out what would really happen instead of depending on studies based on computer models for such vital information."

In 1960, fresh from college, I joined the staff of the Special Power Excursion Reactor Test (SPERT) at the National Reactor Testing Station in Idaho. We were, in a sense, nuclear hot-rodders, pushing scale-model reactor systems to their limits. It seemed at the time to have the flavor of the early days of the Manhattan Project.

In the 1950s and 1960s, our main concern was not the meltdown, but rather a runaway chain reaction, called the "reactivity," or "criticality," accident. In some of the experiments we ran on the SPERT reactor we deliberately withdrew the control rods rapidly from the core. Without the control rods to absorb and regulate the neutrons from the fission process, the chain reaction would spin quickly out of control, and power levels would rise from zero to more than 30,000 megawatts (30 billion watts) in less than one-hundredth of a second. The cooling water would boil furiously, causing a steam explosion.* On one occasion in 1962 I had the dubious distinction of deliberately blowing up the SPERT I reactor.

By the mid-1960s, however, our experiments made clear to us that the new, larger plants being built were effectively safe against these reactivity or criticality accidents.

---

* Steam explosions are somewhat like the explosions that can occur in an ordinary steam boiler. Nuclear explosions, like the one at Hiroshima, are impossible in either a SPERT test reactor or a commercial light-water reactor.

HOW REACTORS WORK

Can nuclear power plants explode like atomic bombs? The answer is no, because commercial U.S. reactors use uranium that is only 3 percent enriched, which is too poor in U-235 to allow for an explosion, even a deliberate one. Instead nuclear power plants release the energy stored in the uranium atom in a slow, controlled fashion.

The heart of a nuclear power plant is the reactor core, which consists of 100 tons of uranium stored in twelve-foot rods stacked vertically. Energy from nuclear reactors can be regulated by inserting control rods into the uranium core. Made of cadmium or boron compounds, these control rods absorb neutrons and can quench the chain reaction. The slightest problem in a reactor will cause the control rods to plunge automatically into the uranium core at high speeds (this is called *scramming* the reactor) and stop the chain reaction.

To carry away the enormous heat generated by the uranium core, water or some other coolant must be circulated continually around the core. Much as coal is burned to produce electricity, the fissioning uranium produces heat which converts a constant stream of cooling water into steam, which turns the blades of a turbine, spinning its blades at enormous velocities. The energy of the spinning turbine blades is then converted into electricity by a process called induction, when a metal wheel turns through a magnetic field (in much the same way that the old bicycle headlamps were powered by the motion of a spinning bicycle wheel). The steam is recondensed back into water and recirculated around the core.

Two types of reactors dominate the commercial reactor scene: the boiling-water reactor (BWR) and the pressurized-water reactor (PWR). Each type cools the core with ordinary water—which is why they are called light-water reactors (LWR)—but in different fashions. Of the 72 operating commercial reactors licensed in 1982 in the United States, all are PWRs or BWRs.

We then directed our attention to the loss-of-coolant accident (LOCA), which could occur if one of the pipes carrying cooling water to the reactor ruptured, and possibly cause a meltdown. Although the reactor would shut down and the fission process would be quenched by the control rods, radioactive decay generates sufficient heat to melt the core if it is not continually bathed in cooling water. (Even after the control rods are fully inserted the core still generates 5–10 percent of its original power, which is sufficient to raise the temperature of the uranium beyond its melting point of 5000°F unless emergency cooling water is dumped onto the core.) In 1963 a new program, an offshoot of the SPERT project, was funded to study these LOCAs: the Loss of Fluid Test reactor (LOFT).

Unfortunately one of the reactor safety programs that suffered the most under

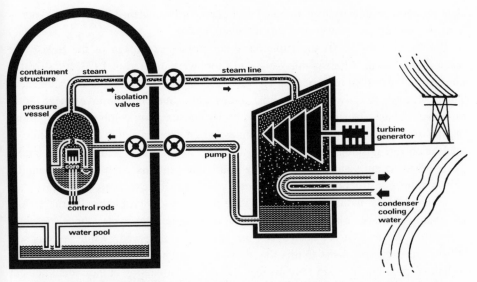

Fig. 1–1   Boiling-Water Reactor (BWR). As with the boiler in a coal-, oil-, or gas-fueled power plant, a nuclear reactor produces steam to drive a turbine which turns an electric generator. But instead of burning fossil fuel, a nuclear reactor fissions slightly enriched uranium to produce heat to make the steam. Water is pumped through the core, boils, and creates steam that is piped to the turbine. [Courtesy Atomic Industrial Forum, Inc.]

Shaw's new management at the AEC was the crucial LOFT reactor. Bogged down by program changes and massive bureaucratic paperwork, the LOFT program took an incredible thirteen years to complete—much longer than it took for a successful lunar landing. Although originally designed to test an actual core meltdown, LOFT was modified to test computer programs predicting the effectiveness of the emergency core cooling system (ECCS) on a scale-model reactor. At the time I left the SPERT project and the nuclear field in 1969, vital reactor safety research had come to a gradual halt.

The 1970s: Growing Doubts

The 1970s saw the continued expansion of this rapidly developing technology. One utility survey indicated that the use of nuclear power rather than coal had saved customers more than $750 million in their 1974 electric bills. The light-water design, only on the drawing boards in the 1950s, completely dominated the world market in the 1970s and was a bonanza for American prestige, technology, and industry.

In 1979 there were 72 licensed reactors operating in the U.S. generating approximately 12 percent of this country's electricity, and 88 more under construction. Worldwide, 530 reactors were in operation. There was good reason for such expansion: the nuclear industry had compiled a safety record unmatched in the entire energy industry, and nuclear power was still relatively cheap.

But by the late 1970s the euphoria about further expansion of the light-water reactor was fading; whereas thirty-eight reactors were ordered in 1973, by 1978 the number had dwindled to two, with eleven cancellations. Long before the accident at Three Mile Island, powerful economic forces were squeezing the investment return on nuclear plants. Declining rates of electricity consumption combined with galloping inflation and sharply higher interest rates to cast a long shadow over the future of nuclear power, pushing the price tag of a reactor to more than $1 billion. Changing government safety standards and costly court interventions contributed to lengthy delays in the construction time for reactors, which already exceeded ten years.

Meanwhile the industry was beset with a new series of problems. By the 1970s, for example, it was obvious that the back end of the fuel cycle, the disposal of radioactive waste, had not been given the attention it deserved. While various scientists and engineers insist to this day that the disposal of waste is a political, not technical, problem, the fact remains that waste is accumulating at individual nuclear plant sites because there is no federal repository for high-level waste. In the next decade the government will probably establish a demonstration waste storage facility, perhaps in Carlsbad, New Mexico, but meanwhile scores of municipalities, counties, and states are passing laws restricting the storage and shipment of nuclear waste within and across their borders.

By 1974 it was also apparent that the AEC had done a better job of promoting than regulating nuclear power. Because of this conflict of interest, Congress abolished the AEC and formed the Energy Research and Development Administration (ERDA, which later merged into the Department of Energy) to study and promote all forms of energy, and the Nuclear Regulatory Commission (NRC) to regulate nuclear power.

To allay growing fears regarding nuclear safety, in 1974 the federal government released a mammoth draft version of a three-year, $3-million reactor safety study led by Dr. Norman Rasmussen of MIT. Using sophisticated methods developed by NASA for its early space program, the Rasmussen report analyzed the reliability of the safety systems in a reactor and the probability of nuclear accidents, concluding that a meltdown was an extremely unlikely event, no more probable than a meteorite's hitting a metropolitan area. Although critics of the report included the prestigious American Physical Society, the Rasmussen report is still widely regarded as a milestone in nuclear reactor safety research.

In March 1975, within a year of the issuance of the controversial Rasmussen report, a fire occurred at the Browns Ferry Nuclear Power Station near Decatur, Alabama. Caused by a technician searching for air leaks with a candle, the fire destroyed several thousand cables, crippled the emergency core cooling systems, and cost the owners over $10 million. Fortunately a last-minute jury-rigged pump allowed continuous flow of coolant above the core and a meltdown was averted.

Just four years later the industry's reputation was dealt another blow. On March 28, 1979, a fifteen-cent part in a valve malfunctioned in a reactor in Pennsylvania and triggered the most severe accident in the history of commercial nuclear power.

At four o'clock that morning outside the city of Harrisburg, Pennsylvania, the Three Mile Island reactor registered a "transient," which is shop-talk among engineers for a glitch. It was nothing unusual; hundreds of small transients are registered at nuclear plants each year without the slighest danger to the public's safety. But this transient was different. The reactor began to heat up and the steam pressure in the pipes began to rise. Finally the pressurizer valve blew open, just as it was supposed to, right on schedule, to relieve excess steam pressure. However, this time the pressurizer valve *stuck open*. As a result, several hundred thousand gallons of cooling water bled out of the pressurizer over the next few hours before the control room operators fully realized what was happening, with up to 75 percent of the core exposed for more than an hour.

The accident at Three Mile Island caught the world unaware. It took a cadre of scientists and engineers months to determine precisely what had happened. The control room operators, high school graduates barely out of reactor school, were ill-equipped to deal with an accident of this type. The Nuclear Regulatory Commission appeared to be bureaucracy at its worst.

In the wake of Browns Ferry and TMI, a flurry of NRC safety regulations were released in order to prevent similar accidents and to improve the performance of reactors. In general the decade of the 1970s was a time when tougher safety standards were imposed on all reactors; improved training of plant operators, tighter fire safety codes, better emergency planning, and more rigorous inspection procedures have reduced the chances of another major accident.

In the 1970s nuclear power advocates also gained an unexpected ally: the energy crisis. Although the oil embargo of 1973–1974 disrupted only 1½ percent of the country's total energy for four months, it knocked the country off its feet for two years. Energy officials declared that new oil supplies—from Mexico, China, or domestic caches—could prove to be illusory. Development of the synthetic-fuels industry would require an $88-billion investment. Nuclear power seemed increasingly attractive in a world where oil supplies were clearly limited and subject to political uncertainties.

The 1980s: Decade of Decision

What lies ahead in the 1980s? One decisive factor seems to be the changing mood of the federal government with regard to nuclear power. In the early 1980s the federal government, convinced that the nuclear industry's problems stemmed from cumbersome regulations, gave the industry its vote of confidence by offering to streamline the regulatory process, to pave the way for a permanent site for waste disposal, and to appropriate sizeable increases in the breeder reactor budget.

But problems remain; lengthy though they may be, licensing reviews still perform a useful function in some cases. For example, just days before the Diablo Canyon plant in California, located near an active offshore earthquake fault line, was to be licensed in 1981, the NRC discovered several crucial construction flaws: not only were the floor supports inadequately reinforced against seismic stress, but due to an accidental reversing of the blueprints for Units 1 and 2, the cable trays and pipe hangers were improperly installed.

Although nuclear power may eventually prove itself to be a safe technology, my concern is that the significance of an accident such as Three Mile Island will be lost in a blizzard of public relations and federal mandates and promises of better safety. I have often wondered: How do you build the perfect machine? For it takes nearly that to contain and control the fission process safely. How do you guard against the inept electrician, the indifferent draftsman, the harried engineer? In the end every nuclear accident can be traced to human error since it is humans who design, build, and operate nuclear reactors. At least, unlike the early years when the few renegade critics had to substantiate all claims that nuclear reactors were flawed in any way, today the burden is on the nuclear establishment. It is up to them to answer justified criticisms—of faulty welds, incompetent operators, and a nonexistent waste disposal program—and demonstrate to the public that the benefits of nuclear reactors are worth the risk.

2

Radiation

Introduction

In the early 1950s it was hard to resist the lure of nuclear power. Not only would it be "too cheap to meter," a catchy slogan that aptly expressed industry and government's great expectations for the burgeoning atomic industry, but there was also the promise that nuclear power would be environmentally superior to other conventional energy sources: it would be clean. Unlike coal-fired power plants, no black smoke would belch forth to settle in the lungs of the nearby population or pollute distant lakes with acid rain. Nuclear power, unleashing the energy inside the atom, was an idea whose time had come.

Thirty years later, hopes for a clean and cloudless nuclear-powered future have been greatly dimmed by the growing fear of radiation contamination. Public trust has been eroded by conflicting "expert" opinions and such notable mistakes as the dumping of low-level waste into the oceans in now-corroding canisters. GIs who in the 1950s observed atomic tests in the Nevada desert were told by the Atomic Energy Commission that "there is no danger" from radioactive fallout. Now they have filed claims against the U.S. government arguing that exposure to radiation during the tests caused them to contract cancer. Nearby residents who woke their children at dawn to watch the mushroom clouds rise on the horizon have filed similar claims.

Although health officials predicted that the radioactive gas vented from the stacks of the crippled reactor at Three Mile Island was relatively harmless, public uncertainty remains. Just how dangerous is radiation?

According to a full-page advertisement in the *New York Times* sponsored by Edison Electric Institute (a consortium of utility companies), we are exposed to more radiation in our living rooms than from nuclear power plants. This surprising statement merits examination and the comparison of various sources of exposure.

Man-made sources of radiation, such as fallout from nuclear weapons, diagnostic x-rays, and radioactive gases released routinely from nuclear power plants, constitute less than half the radiation to which we are exposed; nature supplies the rest. This *natural,* or background, radiation includes cosmic rays from outer space, uranium and thorium in the ground (which means that brick, stone, and other building materials are slightly radioactive), and radioactive chemicals (like potassium-40) found in our bodies. We measure these various levels of exposure with the *rem,* a measure of the biological damage to the tissue caused by a certain amount of radiation. A millirem is 1/1000 of a rem (1 rem equals 1000 millirems).

By way of background radiation the population is exposed to approximately 100 millirems per year, while man-made radiation exposes us to an additional 80 millirems annually. Individual sources of exposure vary: chest x-rays expose the lungs to approximately 10 millirems (0.01 rems), whereas the natural background radiation in a high-altitude city like Denver, Colorado, exposes the local population to 200 millirems (0.2 rems) yearly. By contrast, radioactive gases vented regularly from nuclear power plants on the average account for an exposure of less than 1 millirem (0.001 rem) each year.

The effects of low-level radiation are elusive and hard to prove. It is extremely difficult, if not impossible, to say with certainty that a particular radioactive element was responsible for someone's death. Cancer may not appear for decades after exposure and genetic defects for generations, and even then effects caused by radiation are hard to distinguish from those caused by other environmental hazards like pesticides, food additives, and chemical pollutants. Few low-level epidemiological studies exist, not only because the results are ambiguous, but also because studies that require research and follow-up for a period of forty years are prohibitively expensive. The fact is that *no one knows* precisely the effects of low-level radiation over long periods of time; as a result, the scientific community is polarized in a raging debate.

With little low-dose information available, researchers have concentrated on studies of people exposed to high doses* of radiation, in order to determine the effects of both high- and low-level exposures. Since 1945 when the U.S. exploded the atomic bomb over the Japanese city of Hiroshima, the world has witnessed the devasting effects of radiation. The Atomic Bomb Casualty Commission in Japan compiled an ongoing exhaustive study of 284,000 survivors who were monitored for radiation effects from 1945 to the present. Another large epidemiological study is derived from 14,300 British patients during the 1930s, 1940s, and 1950s who were treated with massive doses of x-rays (up to 2750 rems) for an arthritis of the spine (ankylosing spondylitis). Although other epidemiological studies have been conducted, the Hiroshima-Nagasaki and British x-ray patient data provide scientists with the bulk of information for a wide range of doses.

Today, scientists are in relative agreement as to the effects of *high* doses over short periods of time: 1000 rems to the entire body would kill within a few days to a few weeks after exposure; 500 rems would kill perhaps half of the exposed population within a few weeks; 200–400 rems would cause hemorrhaging, radiation sickness, vomiting, nausea, and fever. Exposure below 50 rems (50,000 millirems), however, would produce no *immediate visible effects* (remember that chest x-rays expose people to approximately 0.01 rems).

---

* A low dose is approximately several rems or less; a high dose is approximately 100 rems or more.

HOW RADIATION WORKS

Although scientists now recognize that there is no "threshold" dose below which radiation is absolutely safe, we do know that trillions of rays can pass through the skin or hit the cells of the body and cause no permanent damage. The potential danger is that some of these rays will disrupt the delicate chemical processes in the cells' nuclei, where genes govern genetic traits. If the rays collide with the atoms in the cell, they could rip electrons from the atoms' shells or disrupt molecules in the cell, thus disturbing the electrical balance between positive and negative charges and creating charged particles—ions. (By contrast, radio, radar, television waves, and sunlight are called nonionizing radiation because the rays are too weak to knock the electrons from the shell of the atom.)

These ions can potentially wreak havoc in the DNA molecules that make up our genes and determine our genetic traits. The chances are rare, but if the ions destroyed the cell's ability to control its own reproduction, and caused it to multiply too rapidly, cancer would result. If these rays passed through the ovaries or the testes, a minute fraction of the ions created in the rays' wake could damage the genes and cause mongolism, cleft palate, or other mutations in offspring.

There are at least three major forms of radiation: alpha, beta, and gamma rays. *Alpha rays* (emitted by many transuranic [heavier than uranium] elements, like plutonium) are capable of traveling only a few centimeters in air and can be stopped by tissue paper. They consist of two neutrons and two protons. *Beta rays* are nothing but electrons emitted from the nuclei of many fission products. They can travel a few feet in air but can usually be stopped by clothing. *Gamma rays* are a form of electromagnetic radiation (like light, radio, and television) with very short frequencies. Energetic gamma rays can penetrate several inches of steel. Per unit of energy, the biological damage is about the same for beta and gamma rays, while alpha rays cause ten to twenty times more damage.

Since radiation effects are not readily observable below 50 rems, some people wonder how the effects of low-level radiation can be determined at all. Very simply, scientists study the high-dose effects and extrapolate the information down to the low-dose range. Here lies the source of the radiation controversy. Although pro- and antinuclear scientists draw on the same data—studies of the Hiroshima-Nagasaki casualties and British x-ray patients—they interpret the information differently (see Fig. 2–1). Until recently most scientists accepted the *linear hypothesis,* by which it is assumed that the damage caused per rem is the same at low doses as at high doses (i.e., doubling the dose doubles the damage). Increasingly, however, some scientists

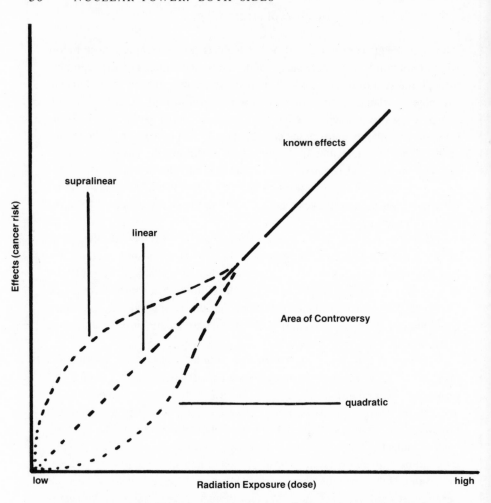

Fig. 2–1   Three Hypotheses to Determine the Effects of Low-Level Radiation. The solid line represents the well-established cancer risk from high levels of radiation. Note that as the exposure increases, the number of cancers rises linearly. The effects of *low* levels of radiation, represented by the dotted lines, are not as well established, and are estimated by one of three hypotheses: the *linear hypothesis,* which assumes that the cancer risk at low levels is the same as at high levels; the *quadratic hypothesis,* which assumes that less damage is caused per rem at low doses than at high doses; and the *supralinear hypothesis,* which assumes that radiation causes more damage per rem at low doses than at high doses. All three dotted lines arrive at zero effects when the radiation exposure is zero.

think that linearity may even overestimate the risk at low levels. They support the *quadratic hypothesis,* by which it is assumed that *less* damage is caused per rem at low doses than at high doses because evidence suggests that the body has the ability to repair cells damaged by low-level radiation. A small minority of scientists supports the *supralinear hypothesis,* by which it is assumed that *more* damage is caused per

WHO SETS THE STANDARDS?

In the 1920s, as the new x-ray technology flourished, it became apparent that more and more radiologists and technicians were developing skin cancer and possibly leukemia. As a result, two nongovernmental organizations of scientists were formed to provide guidelines for occupational exposure to radiation:

1. The *International Commission on Radiological Protection (ICRP)* was born out of a gathering of radiologists in 1928 at the Second International Congress of Radiology in Stockholm. Formerly the International X-ray and Radium Protection Commission, this group is comprised of one chairman and twelve members who are chosen from the fields of radiology, radiation protection, physics, biology, genetics, biochemistry, and biophysics, without regard to nationality. The ICRP meets every three years (interrupted only by World War II), and nominates new members.

2. The *National Council on Radiation Protection and Measurement (NCRP)* is the U.S. counterpart to the ICRP, formed in 1929. The NCRP consists of 75 members, elected similarly to the ICRP, and 350 additional scientists who serve on various NCRP committees. NCRP members, who are often U.S. representatives to the ICRP, work closely with the ICRP to make recommendations to the U.S. government.

Following World War II, as more nations acquired nuclear technology, another group was established:

3. The *United Nations Scientific Committee on the Effects of Atomic Radiation (UNSCEAR)* includes member scientists appointed by their respective governments.

Although the recommended exposures put forth by these three organizations do not carry the force of law, they are usually adopted by governments and industry throughout the world. In the U.S., maximum limits of exposure have been established at:

- 0.5 rems (500 millirems) per year to individual members of the public;
- 5 rems (5000 millirems) to atomic workers in universities, government laboratories, and the nuclear industry; and
- a ceiling of 0.17 rems (170 millirems) as an average yearly dose to any population group.

These numbers were determined by analyzing hundreds of radiation studies, such as the Hiroshima-Nagasaki and British x-ray patient data.

rem at low doses than at high doses. These scientists theorize that perhaps low dose weaken and damage cells (which live on to damage other healthy cells), whereas hig doses simply kill cells.

Divergent Views

Pro-nuclear scientists outnumber the nuclear critics. They point out that radiation perhaps the best understood environmental hazard—better understood than air po lution, insecticides, or food additives. They argue that radiation is a relatively simp phenomenon (see "How Radiation Works"), and that many studies have been co ducted on radiation effects.

Scientists throughout the world, in different laboratories, at different times, ha examined different patients from different races, and have drawn conclusions that a largely consistent. Their conclusions are supported by the *standard-setting bodies-* prestigious international committees of experts—who continually evaluate the da and recommend permissible limits of exposure that form the standards for gover ments and industries throughout the world (see "Who Sets the Standards?").

Those who support the pro-nuclear scientists also call attention to the health ris associated with other energy sources, such as coal or oil. "Nuclear power certain involves some risk," said Floyd W. Lewis, chairman of Middle South Utilities, Inc "but I know of no other high-technology industry which can point to the most serio accident in its history and say truthfully that no one was killed or injured." T advocates ask that harmful effects of nuclear power be kept in perspective, and judg quantitatively. When the number of deaths calculated from nuclear power as an ener source is compared with that from coal pollution, advocates are confident that nucle power provides the better option.

Nuclear critics maintain that the cumulative effects of natural and man-ma radiation, combined with inadequate monitoring at nuclear facilities and in uraniu mines, make radiation exposures more dangerous than the government or indust would have us believe. Their sentiments are perhaps best expressed by Alb Schweitzer, who upon learning that experts had determined permissible doses, aske "Who permitted them to permit?"

Antinuclear scientists also believe that the effects of low-level radiation have be underestimated. In the 1950s Dr. Linus Pauling, winner of two Nobel Prizes, warn about possible genetic defects caused by exposure to fallout. In 1969 Dr. John Gofman, then an associate director of the AEC's largest laboratory for the design nuclear weapons, in Livermore, California, using the linear hypothesis, announc that if everyone in the United States were exposed to the allowable limit (0.17 re in a given year), up to 32,000 extra cancers would result.* Other scientists ha

*From Gofman's calculation we can deduce that 1000 person-rems would cause one fatal cancer person-rem is a measure of the amount of radiation absorbed by a given population. If 1000 peo

warned that some of our most comprehensive studies may contain hidden biases that prejudice the results. In addition, controversial low-level radiation studies conducted n atomic workers within the last decade indicate that certain forms of cancer caused y radiation exposure may be up to twenty-five times more prevalent than suspected.

In 1977 Dr. Thomas Mancuso of the University of Pittsburgh published the largest ow-level epidemiological study to date, in which he linked workers' radiation expo-ure to an unusually high number of cancer deaths at a large government facility in Hanford, Washington. Funded by the AEC, Mancuso analyzed the records of 24,939 workers at the Hanford Atomic Works from 1944 to 1977; unlike the Hiroshima-Nagasaki casualties, the employees wore badges monitoring their exposures. In 1976, year before Mancuso announced his results, the government, after twelve years of unding and more than $6 million in government grants, terminated his contract ecause he was publishing at an "inordinately slow rate." The standard-setting bod-es rejected the validity of the Mancuso study on the basis of faulty methodology, which is crucial to the interpretation of any data. But antinuclear scientists were not onvinced; they claimed that Mancuso's funding dried up because the government did ot like what he was reporting.

The Political Arena

Whether or not Mancuso's findings are valid, his and other low-level studies under-ne an important feature of the debate: enough uncertainties exist in the data that cientists cannot precisely predict the long-term effects of low-level radiation. Our neans of extrapolating information from the Hiroshima-Nagasaki data and from other udies to determine the effects of Three Mile Island—by choosing one of three ypotheses—are precisely what the scientific term implies: *hypothetical*.

Given the polarity of the debate and the fact that the studies are sparse and contro-ersial, one simple statement—that the majority of scientists think that current per-issible exposure limits are adequate—is rarely accepted at face value. Antinuclear cientists believe that the majority of scientists, subject to political pressure from owerful government laboratories and the nuclear industry, cannot be depended on o arbitrate the more sensitive issues, because more than once the majority has been ccused of subtly defending the status quo. Pro-nuclear scientists insist that the majority pinion has evolved naturally from scientists conducting research independently hroughout the world, and that dissenting scientists exploit the uncertainties in the ata, prey upon public fear dating back to Hiroshima, and exaggerate the dangers of ommercial nuclear power. These conflicting opinions frequently enter the radiation ebate.

---

ch received 1 rem of radiation, or if 500 people each received 2 rems of radiation, the population in ach case has received 1000 person-rems. The standard-setting bodies assume that 5000–10,000 person-ms would produce one fatal cancer, whereas critics of these standards believe that anywhere from 100 1700 person-rems would produce a fatal cancer.

Antinuclear scientists would challenge the credibility of a majority opinion. "With an issue as sensitive as radiation effects," said one scientist, "you cannot afford to be ruled by a majority vote as though you were running a popularity contest." Dissenting scientists complain that the government's near-monopoly on health research at laboratories and universities has all but squelched independent analysis of data. One reply to these allegations would be that independent groups of scientists also research radiation effects. However, antinuclear activists point out that the impartiality of one prestigious independent scientific body, the National Academy of Sciences, has been called into question more than once. In 1980, for example, after a NAS study was published concluding that cholesterol intake was not linked to heart disease, many people were distressed to learn that the scientists who issued the study had ties to the food, dairy, and egg industries. "They [at the Academy] do not reflect dispassionate analysis," said Dr. Henry Kendall, chairman of the Union of Concerned Scientists. "There's a consistent pattern of having the goat guard the cabbage patch."

To the pro-nuclear scientists, the dissenting minority consists of formerly reputable scientists who have been lured by fame and publicity to break from their solitary research and plunge into the media spotlight. Pro-nuclear scientists maintain that antinuclear scientists make shrill, alarming statements based on scant data, and are disturbed that the media feed upon the antinuclear activists' sensational claims. (As an example, when a British investigative team learned that John Wayne, who died of cancer, filmed the movie *The Conquerors* in the Utah desert near a fallout test site, several magazines reported that John Wayne died from exposure to fallout.)

These divergent views have immediate implications for our health. In this chapter, four scientists who have shaped our understanding of the effects of radiation investigate the controversy. All four base their scientific opinions on the Hiroshima-Nagasaki and British x-ray patient data, but arrive at different conclusions. In the first two essays two health physicists dispute our chances of getting cancer and genetic defects from exposure to radiation from neighboring nuclear power plants. But the radiation debate goes even further; it raises the profound question of whom we can trust in this country to give us the truth about scientific issues. The last two essays of this chapter discuss the explosive political implications of the debate: Is the scientific establishment lying to the public about the dangers of low-level radiation or are the few dissenting scientists crying wolf? This question and others are answered in this chapter by scientists on the forefront of the nuclear debate.

Underestimating the Risks

KARL Z. MORGAN

Karl Z. Morgan, former director of the Health Physics Division at the Oak Ridge National Laboratory in Tennessee, is currently a professor of nuclear engineering at the Georgia Institute of Technology. Often referred to as the "father of health physics," Dr. Morgan was the first editor of *Health Physics* magazine and was awarded the first gold medal from the Swedish Academy of Science in 1962. For twenty-five years he was chairman of the dosimetry committees of the ICRP and the NCRP. Dr. Morgan now warns that safety standards with respect to radiation may be as much as ten times too lax.

In 1943, at Oak Ridge, Tennessee, where uranium for the world's first atomic bomb was being enriched, radioactive liquid waste was being dumped into the nearby White Oak Lake. As director of the Health Physics Division at the Oak Ridge National Laboratory, I was responsible for determining the amount of waste that could safely be put into the lake. The only standard I had to go by was the National Council on Radiation Protection (NCRP) value of 0.1 R* per day for workers. Because there was a war going on, some of my engineering associates at the lab were provoked at my conservatism. They insisted that I should raise the level of the standard a thousand times higher, to 100 R per day, since no one would drink or swim in the water because signs and fences discouraged fishing and recreation in the lake area. They also pointed out that the waste in the lake would be diluted when it later emptied into the nearby Clinch River.

Forty years later we now recognize that my associates' recommendation would have been over seven million times the value we would consider acceptable today. I stuck to my guns and used the lower figure of 0.1 R per day because I was concerned that the radioactive waste could be reconcentrated in the ecosystem. I deserve no special praise, however, because in the context of today's standards, my recommendations were still *7000 times too high*. Years later my fears were confirmed when Dr. H. M. Parker, my close associate at the Hanford Atomic Works in Washington, reported that white fish swimming on the bottom of the nearby Columbia River were concentrating phosphorus-32 in some of their body tissues by a factor of a million.

This was not an isolated incident. Over the past thirty years several compelling studies on atomic workers and children exposed to x-rays have called into question the adequacy of our radiation standards and have forced us to reevaluate earlier studies, such as the Hiroshima-Nagasaki data. During my tenure as chairman of the Internal Dose Committees of both the ICRP and the NCRP, I approved limits of radiation exposure to the various radionuclides based on the best scientific evidence available at the time, which by today's standards would be considered dangerous. As the latent

R refers to roentgen, a measurement of ionization, which is comparable to approximately 0.88 rem.

effects of cancer have begun to reveal themselves, health physicists like myself have
slowly changed their thinking[1] and now realize that low-level radiation poses a greater
health risk than the nuclear industry and the standard-setting bodies would have us
believe.

The Early Years

We have nothing to be proud of, looking back at the development of our understanding of radiation. As far back as 1500 workers in the cobalt mines of Saxony and
pitchblend mines of Bohemia were afflicted with a mysterious "mountain sickness."
Their high death rate was due to radioactive radon gas which seeped out of the uranium contained in the pitchblend ore. Without proper ventilation of the mines, the
alpha particles from this odorless, colorless gas and its daughter products bombarded
the lung tissue of the miners and eventually caused lung cancer. It was of course not
known that this radioactive by-product of the decay of uranium caused their deaths
for it was not until 1896 that Becquerel discovered radioactivity. It is a sad commen-
tary on our society that it took 450 years for man to learn such a simple lesson about
radiation, but I find it even more disturbing that hundreds of years later, uranium
miners in the United States are still dying from lung cancer from exposure to radon
gas in the mines.

Folklore played a large role in our ignorance of the subject. As late as the 1920
radiation was thought to be a cure-all, and was used to treat everything from acne to
insanity. Water and wines with high levels of radioactive radon gas were widely
advertised and sought for their purported healing powers. One textbook that was used
to train x-ray technologists even stated that Negroes should be given more exposure
because of the thickness of their skin!

Hidden Biases

Everything changed in 1945, when the atomic age was ushered in dramatically by
the holocaust of Hiroshima and Nagasaki. Even today the data from several hundred
thousand atomic bomb casualties serves as one of the landmark studies on the health
effects of radiation on humans. Unfortunately they are unqualifiedly accepted without
calling attention to or correcting some serious, glaring biases that underestimate the
cancer risk.

One serious bias in the case of the Hiroshima-Nagasaki casualties might be that
those who survived the bombings were slightly stronger and healthier than average
and naturally would be less susceptible to cancer. The Japanese were exposed not
only to ionizing radiation but to blast, fire, deprivation, and the aftermath of
momentous tragedy which weakened their immune systems. Many deaths occurred

om common diseases for several months following the bombings. Thus those who ere weakest and most likely to die of cancer died early of pneumonia and other seases without surviving long enough to die of cancer and be counted in the studies. he tragedy and human suffering at Hiroshima and Nagasaki may have selected out e healthier Japanese, those less likely to get cancer, and produced many *in situ* nrecognized) cancers in persons who died early from consequences of the bombing fore their malignancies could be diagnosed and identified.

This speculation is further substantiated by the work of Dr. Joseph Rotblat, pro-ssor of physics emeritus at the University of London.[2] He reexamined the atomic sualty data and studied the leukemia rate among those who entered Hiroshima ithin the first three days after the explosion to look for friends and relatives and ere thus exposed to the radioactive contamination from fallout and residual neutron-duced radioactivity. Rotblat found that the early entrants were ten times more likely die of leukemia than one would expect from other leukemia death rates quoted by e standard-setting bodies (NCRP, UNSCEAR, and ICRP). His data strongly indi-te that the atomic casualty data underestimate the risk of cancer if we examine pulations that were not subjected to the enormous human suffering and disease und at Hiroshima.

The same biases can possibly be found among the British x-ray patients suffering om an arthritis of the spine (ankylosing spondylitis). In particular, Dr. Edward P. adford, an epidemiologist at the University of Pittsburgh, and a group of scientists examined the medical records and found that this population was more likely than e average to suffer from common diseases such as tuberculosis, chronic bronchitis, neumonia, influenza, and cerebrovascular disease. Like the Hiroshima-Nagasaki sualties, many of these patients probably did not live through the normal latency riod of malignancies.

Another hidden bias in the Hiroshima-Nagasaki data concerns the rather low inci-nce of cancer among the children who were exposed to the radiation from the bomb hile still in the uterus of the mother. This low incidence of cancer can be explained y the high rate of spontaneous abortions following the explosion.[3] In other words, e fetuses that were more likely to have developed cases of radiation-induced leu-emia received such high doses and were subjected to so much trauma that they failed eferentially to survive.

This speculation seems to be substantiated by one of the earliest studies on the arcinogenic effects of prenatal exposure to radiation from x-rays. Dr. Alice M. tewart of Oxford University conducted a massive follow-up of some twenty million ve births in the United Kingdom from 1943 to 1965, among which there were about 3,000 cancer deaths before the children reached the age of ten. She found that, on e average, 1700 person-rems* resulted in one cancer death for fetal doses ranging

A measure of the amount of radiation absorbed by a given population. For a more detailed explanation, e pp. 32–33.

between 200 and 460 millirems, which was a much lower dose than that expected from previous animal studies.

Stewart's study originally received some criticism that her control population was inadequate. But studies have since confirmed her results and a number of publications have indicated that if any uncontrolled biases could be accounted for, the cancer risk from prenatal exposure would tend to increase rather than decrease. Today the Stewart results are almost universally accepted, and her figure of 1700 person-rems per cancer death probably should be considered as a minimum estimate of danger from prenatal exposure rather than the average estimate.

Historically the Hiroshima bomb data have been questioned by health physicists for yet another reason: they showed that radiation is less damaging than what one would expect, judging from the bulk of other epidemiological studies. But because of the sheer volume of the data from the Hiroshima bomb blast, these results tended to overshadow all other studies when setting radiation standards. This contradiction was resolved in 1981 when scientists at the Lawrence Livermore National Laboratory showed that there were significant errors in the original calculation of the neutron and gamma dose from the bomb, as much as ten times in certain cases. In other words, the Hiroshima data have been fundamentally flawed for thirty-five years. Although the final results from these revisions will not be settled for years, it is now obvious that radiation may be several times more dangerous in causing certain forms of cancer than previously thought.

New Studies in the Last Two Decades

Not only must we reexamine our previous reliance on early data, we must also seriously consider the implications of recent studies that directly challenge the radiation levels set forth by the standard-setting bodies.

Consider, for example, new studies on the ratio of genetic mutations to cancers. Fifteen years ago the scientific community was rather smug, as some scientists still are today, in believing that genetic risk was much larger than cancer or other somatic risks (such as shortening of the life span or premature aging), and that somatic risk was almost negligible at low doses. Today, however, even the ICRP admits that "it could be concluded that the ratio of somatic to genetic effects after a given exposure is 60 times greater than was thought 15 years ago."

Genetic Risks   Genetic mutations caused by radiation are notoriously hard to study and interpret because of the long time lapse between generations. Furthermore, a mutation may be dominant and show up in the very next generation, or it may be recessive and nonvisible and not appear for several generations, if at all.

I wish to sound a warning that I am sure my longtime friend, the geneticist and Nobel laureate H. J. Muller, would urge me to make were he still alive: it may well be that in the long run these recessive mutations, which result in a lack of vigor, susceptibility to disease, and a slight reduction in mentality and physique, will be a far greater burden to society than the easily identifiable dominant mutations. This is because the recessive mutations are eliminated very slowly from the gene pool and may not be recognized for what they are as they gradually mount over many generations.

Muller found that for every visible, dominant mutation there may be 10,000 nonvisible recessive mutations in organisms that appear perfectly normal. The implications are enormous: radiation may be one of the principal etiological agents responsible for human suffering and disease, causing a general lack of resilience, a failure to attain a high degree of perfection, and a general deterioration of the human race. Thus many who are damaged genetically by a medical x-ray procedure or from radioactive effluents from nuclear operations may not be born until generations after a forebearer received the radiation exposure. The questionable ability of a protective society to eliminate unwanted genetic mutations casts doubt on the reliability and acceptability of present radiation standards.

Cancer Risks   In the early 1970s Dr. Irwin Bross[4] and other scientists at the Roswell Park Memorial Research Institute in Buffalo, New York, examined children who received diagnostic prenatal x-ray exposure, some of whom later developed allergic diseases such as asthma. Bross found that children aged one to four who had prenatal exposure only, but not asthma, ran a 50 percent increase in the risk of getting leukemia, which verified Dr. Stewart's earlier study on the increased cancer risk of children exposed prenatally to x-rays. He found a 300–400 percent increase in the leukemia risk for children who suffered from asthma and similar respiratory diseases, but who were not exposed prenatally to x-rays. And he found a *5000 percent* increase in leukemia among children suffering from *both* these diseases and prenatal x-ray exposure. Thus the coupled or synergistic effect of both respiratory diseases and prenatal exposure to x-rays generated leukemias that far exceed the sum of either one or the other considered separately. One important risk from radioactive pollutants that is frequently overlooked is that cancer-causing agents do not necessarily act independently.

Dr. Bross's study suggests that radiation weakens the body's defense system, making the body susceptible to diseases, including cancer. In a life-and-death race between cancer and common diseases, the noncancerous disease may very often win out and be diagnosed as the cause of death, especially in cases of high-level radiation exposure. Thus as we develop cures for common diseases, we can expect the risk from radiation-induced cancer to increase.

Of course the scientists on the standard-setting bodies have tried to refute many of these new studies, including one by Dr. B. Modan of Chaim Sheba Medical Center at Tel Aviv, Israel, and his colleagues.[5] They examined 11,000 immigrant children in Israel who were treated with x-rays to combat ringworm of the scalp and found a

much higher incidence of thyroid cancer than would be expected in a normal popu
lation. According to Modan, as little as 5000 person-rems of radiation would resul
in one thyroid cancer, which is a number much smaller than the figure quoted by th
ICRP or UNSCEAR based on earlier studies with iodine-131.* Some members of th
standard-setting bodies complained and gave explanations why Dr. Modan's figure
were incorrect. They leveled objections at Modan's study by quoting figures from
other studies of radiation exposure to iodine-131, but overlooked the fact that th
doses they quoted were so high that much of the thyroid tissue was destroyed or mad
incapable of reproduction. They failed to recognize that high doses selectively destroy
most of the tissue that may be incubating a cancer, thereby preventing a malignancy
from originating from the (dead) thyroid cells. Certainly they would not have u
believe that thyroid carcinoma could originate from dead thyroid tissue.

The Mancuso Study

Perhaps the most controversial of all the new studies was one carried out by Dr
Thomas Mancuso of the University of Pittsburgh, who along with Drs. Alice Stewar
and George Kneale studied the mortality rates of radiation workers at the Hanfor
Atomic Works.[6] They found that as little as 120–140 person-rems could cause on
cancer death, an estimate considerably lower than the 5000 person-rems quoted b
the standard-setting bodies.

Although there has been much criticism of the Mancuso study, responsible critic
were forced to admit that there was a significant increase in multiple myeloma (
cancer of the bone marrow) and cancer of the pancreas. The Mancuso study als
confirmed the earlier findings of a study conducted by Dr. Samuel Milham, an epide
miologist with the Washington State Public Health Department, who had first calle
attention to the unusually high incidence of cancer at the Hanford Works.[7]

When Dr. Milham first announced his findings in 1974, the Energy Research an
Development Administration (ERDA, later the Department of Energy) urged Man
cuso to publish his data and refute Milham, because a preliminary examination o
Mancuso's data suggested there might be no unusual increase of cancer among th
radiation workers. However, Mancuso refused to rush his findings into publicatior
he wanted to avoid a hasty conclusion, and insisted that he have independent medica
and epidemiological reviews of his findings. Following this, ERDA discontinue
support of Mancuso's program and moved the study in-house, where it was continue
by scientists at the Hanford Atomic Works. There followed many charges and coun

*In medicine, because iodine-131 has the tendency to concentrate heavily in the thyroid gland, it
injected into a patient's body to determine malfunctions in that gland. In a nuclear reactor, iodine-131
cesium-137, and strontium-90 are a few of the hundreds of radioactive fission products contained insid
the core. If these fission products were released into the environment in an accident, scientists believe tha
iodine-131 could pose the greatest cancer threat because (1) it is very plentiful and (2) it is easily dissolve
in water and incorporated in milk.

tercharges as to why ERDA terminated Mancuso's funding. ERDA claimed the move was made because of Mancuso's imminent retirement (he was five years under retirement age) and because he had not shown an interest in expanding his program. But Mancuso did want to expand his study to include the atomic workers at the Los Alamos laboratory in New Mexico, and it was ERDA which turned a deaf ear to his proposals. Many people, myself included, believe the real reason for moving the program in-house was the fear that Mancuso might report findings of an increased cancer incidence related to the recorded radiation exposures. Information brought to light in congressional hearings and through the Freedom of Information Act has lent strong support to this accusation against ERDA and the Department of Energy (DOE). The study is still continuing (though at a lesser pace since the withdrawal of DOE support), and it is interesting to note that recent evaluations and publications of the data by Drs. Mancuso, Stewart, and Kneale have served only to strengthen their earlier findings.[8]

Perhaps the principal criticism of the Mancuso study was that Dr. Mancuso found no significant increase in leukemia among the radiation workers at Hanford. Instead he found an increase in pancreatic cancer and multiple myeloma and an unusually *low* incidence of leukemia, while the Hiroshima-Nagasaki data showed just the opposite. But in light of the many recent reevaluations of the atomic bomb casualty data, perhaps we should ask why the Hiroshima-Nagasaki data and the British x-ray patients study found such a *high* incidence of leukemia, rather than ask why the Mancuso study found such a *low* rate. I should explain. Leukemia seems to dominate over myeloma in cases where people are (1) exposed to high doses of radiation and (2) exposed to blast, fire, deprivation, great mental and physical stress, and various diseases, such as arthritis of the spine and respiratory ailments. Leukemia does not seem to dominate over myeloma when these two conditions are not satisfied. Leukemia dominated over myeloma in the Hiroshima-Nagasaki data, the British x-ray patients, the early entrants into Hiroshima, and the children in the Bross study who developed certain respiratory diseases in early childhood. All these studies fit the two conditions. Populations which did not fit them, such as Stewart's study of prenatal exposure and the Mancuso study, did not show leukemia dominating over myeloma. If our theory is correct that trauma and disease cause leukemia to dominate over myeloma in a population, then the Mancuso data are consistent with the previous data on radiation-induced cancer. These new studies force us to reevaluate the effects of low-dose radiation.[9]

We will never settle the low-level radiation controversy unless government laboratories and universities conduct more large epidemiological studies. All too often the scientists who sit on the standard-setting bodies are subject to conflicts of interest, and government funding is used to place pressure on researchers. Radiation standards should not be determined by a vote of a committee of scientists, nor can standards for people be determined by radiation studies on fruit flies and fungi, or solely by theoretical considerations.

The Industry's Track Record

Nuclear power plants routinely discharge thousands of curies* of tritium (a radioactive isotope of hydrogen) and radioactive noble gases (such as krypton-85 and xenon-133) annually, and many plant operators neglect to measure the discharges of carbon-14, which in some cases is a major contributor to the population dose. These gases are dispersed rapidly into the biosphere, and some scientists believe that the nuclear industry by the year 2000 could increase levels of background radiation by about 3 percent, which would add about 7000 fatal cancers per year to a world population of four billion ($[4 \times 10^9$ people$] \times [6 \times 10^{-4}$ fatal cancers per person-rem$] \times 0.1$ rems per year$] \times 3\% = 7000$ cancers per year).

Ultimately we must translate these figures into human terms. It is little consolation to a mother living near a nuclear power plant to know that the average risk to the people in her community is 1700 person-rems per cancer, when she learns that her child with asthma has fifty times greater risk of developing cancer. It helps very little to tell the mother that natural background radiation is approximately 100 millirems per year and that radiation from the plant is only 5 percent of this or 5 millirems per year. She probably is very unhappy to learn that this 5 millirems per year gives her child with asthma a 0.5 percent risk of dying of cancer after a thirty-year exposure. Nor does it help to tell her that if a coal-burning power plant, even an unusually clean one, were to replace the nuclear power plant, the risk might go up above 0.5 percent and that the primary risk would then become one of chronic bronchitis and emphysema rather than cancer. It is difficult for this mother to understand why she should risk the life of her child so that the power plant can be located at a particular river site or, as she may rationalize, so the stockholders can expect a better return on their investments.

Another concern is the growing practice of "burning out" temporary employees: many nuclear power plants find it necessary to solve the problem of an individual exposure limit for repair work in high radiation-exposure areas by hiring temporary employees to spread the dose on "hot" operations. Workers who need jobs move from plant to plant, where records often are not maintained, and they burn out again and again. Even if each individual exposure were below the maximum permissible, this practice of burning out employees is undesirable because temporary employees usually do not appreciate the long-term risks of radiation exposure and because the total number of person-rems is always greater for untrained workers.

During the mid-1960s the government and industry pledged to try to keep exposures As Low As Reasonably Achievable (ALARA). Unfortunately, many nuclear power plants had already been constructed or designed in an earlier, less conservative

---

*A curie originally corresponded to the radioactivity emitted by 1 gram of radium, but is now defined as the amount of any radioactive substance where 37 billion of its atoms are undergoing radioactive decay per second.

period, and owners of these plants could make an effort only to keep exposures within the maximum radiation limits, but not ALARA. Even today many utility companies and government regulatory agencies are slow to take ALARA seriously.

The operation of many of our nuclear laboratories and installations is nothing to brag about. Laboratories like Oak Ridge in Tennessee, Argonne in Illinois, Brookhaven in New York, and the Savannah River operations in South Carolina have a relatively good record because for decades they have conformed successfully to ALARA and the average dose to workers has been less than 10 percent of the maximum permissible exposure of 5 rems per year. But I was particularly unhappy, for example, with what went on at the West Valley, New York, reprocessing plant* or the Kerr-McGee Oklahoma fuel fabrication plant: leaks, safety violations, overexposures of workers, spills of radioactive materials, etc. The operators at West Valley in particular committed every known mistake in health physics and went on to invent some new ones. I believe in these instances there was wanton disregard of ALARA and of good health physics practices. The eventual closing of the West Valley fuel reprocessing plant left the nearly bankrupt state of New York with a tiger by the tail and a cleanup operation that could cost over $1 billion.

Unfortunately, general negligence in the nuclear industry has led to a rather deplorable situation:

o   Some 200 uranium miners have died of radiation-induced lung cancer because radon levels in the underground mines were too high.

o   Thousands of people have been exposed to radiation from uranium sludge piles left over from the mining operations throughout the Rocky Mountains and Midwest.

o   Hundreds of square miles of Rocky Flats, Colorado, has been contaminated with plutonium that escaped into the air from several large fires and improper storage of plutonium waste at the government's sprawling weapons facility. Now plutonium poses a threat of radiation-induced cancer to hundreds of residents for many hundreds of generations to come.

The Politics of Nuclear Power

Our society is strongly polarized on the question of risk of exposures to low levels of ionizing radiation: the truth probably falls somewhere in between the two extreme points of view. The pro-nuclear people tend to underestimate the risks of low-level radiation, perhaps because they fear that the public might not accept nuclear power if the radiation risks were better known or because they wish to avoid the costs of

*Nuclear waste, because it contains 1 percent plutonium-239 and 1 percent U-235, can be chemically reprocessed to extract any unused fissionable material from the waste fission products. For more information, see Fig. 4–1.

additional measures of radiation protection which they believe are unnecessary. The pro-nuclear group will go to any extreme to sell nuclear energy, exaggerate its merits, depreciate its weaknesses, and underestimate its risks.

On the other hand the antinuclear groups often exaggerate the risks of low-level radiation, especially those risks associated with nuclear energy. They seem to fear 1 millirem of ionizing radiation from a nuclear power plant more than they fear 1000 millirems from unnecessary medical diagnostic procedures. Many antinuclear people exaggerate the seriousness of nuclear power accidents and would slam the door shut on all future nuclear power development without recognizing the undesirable consequences of remaining alternatives to nuclear energy.

Part of the difficulty is that nuclear energy was introduced to the public during World War II as a weapon of heretofore undreamed-of destructiveness. This of course portends its possible use in the next war, where hundreds of millions will be injured or killed by fire, blast, and radiation.

After World War II a number of big business corporations in the United States, that played such a magnificent role in winning the war, took up the Atoms for Peace slogan: they would turn their swords into plowshares. However, the early development of nuclear energy took place under the cloak of tight security. A type of parental priesthood developed among the scientists, engineers, and members of government agencies such as the old Atomic Energy Commission. In some cases security overstepped its bounds and went so far as to imply that certain people were not patriotic if they suggested that low-level radiation exposure might be harmful to humans.

The electric utility companies were oversold on the advantages of nuclear power. Now, to allay fears and convince the public that nuclear reactors are safe—and to save on our utility bills—the utilities appear, with tongue-in cheek, to downplay the seriousness of low-level radiation. They are joined in this chorus by big industry, national laboratories, government agencies, and members of the standard-setting bodies, most of whom have a stake in the utility industry and make their living by its success.

There is probably no occupation that is free from some risk to its workers. I believe the radiation controversy originally developed because many people in the nuclear industry and federal agencies inadvisably proclaimed that there was *no* radiation risk. If the proponents of nuclear energy had been more reasonable in their claims about radiation safety, they would not now be trying desperately to save face.

Few of the industry's claims, however, are justified. Savings on our utility bills would be greatly reduced or even become a liability if all the costs of research and development, waste disposal, decommissioning of facilities,* and health effects were added to the bill instead of being picked up by the government. Present light-water reactors are very inefficient and wasteful of uranium. These reactors were designed

---

*After thirty to forty years a nuclear power plant must be shut down permanently due to the cumulative effects of radiation, especially radiation damage to the vessel. In the United States no large-scale commercial reactor has yet been decommissioned.

under the old threshold hypothesis and features were not incorporated to make full use of shielding, remotely operated equipment, instrumentation safety devices, etc.

Nuclear power is as much a social and political problem as it is a scientific one, and we must come up with a comprehensive plan of action. First and foremost, I would like to see the radiation standards for workers lowered by a factor of at least two—from 5 rems per year to 2.5 rems per year, which would undoubtedly protect the lives of many workers. A number of citizens' organizations have demanded that maximum radiation standards be lowered by the factor of ten. I do not think that would be wise at the present time for several reasons:

First, if we lowered the maximum permissible levels by a factor of ten, many of our present nuclear power plants could not continue operating. In pressurized-water reactors (PWRs), for example, necessary repairs in the vicinity of the steam generators precludes the lowering of workers' radiation levels by a factor of ten. The high radiation fields near the steam generators are largely caused by cobalt-60. One solution would be to use more radiation shielding, more remotely controlled equipment, and to redesign our PWRs so that certain precursor cobalt radionuclides do not circulate in the primary cooling water or enter high neutron fields.

Second, since 90 percent of man-made radiation exposures arise from medical uses of radiation, it would be foolhardy to reduce radiation levels in nuclear power plants while leaving intact large unnecessary and unproductive exposures from medical x-rays, which are exempt from the recommendations of the ICRP and NCRP. We need to implement stringent monitoring of radiation exposure from *all* sources, including diagnostic equipment.

Third, operators of the plants would simply hire more "jumpers" and burn out more workers. Though each worker would be exposed to less radiation, more workers would be hired to do the same job, so the total number of person-rems, and therefore the number of radiation-induced fatal cancers, might actually *increase*.

Simply reducing levels of the radiation standards by a factor of ten would be an isolated, simplistic solution, much like putting a finger in the hole of a leaking dike. To arrive at a comprehensive plan, I would suggest several other actions:

o   Take measures to reduce the overall number of person-rems of exposures. For example, set a limit each year of 500 person-rems per 1000-megawatt electric nuclear power plant and reduce it to 200 person-rems for plants now on the design boards.

o   Take bold steps to reduce all unnecessary medical exposures, which constitute the bulk of our exposure to man-made ionizing radiation.

o   Apply the philosophy of ALARA everywhere, not the policy of burn-out, to *all* kinds of exposures, including chemical agents like asbestos, pesticides, and food additives, as well as nonionizing radiation such as radio, radar, and television waves. The goal should be to reduce the number of malignancies and genetic defects regardless of the causative agent.

○   Give adequate support to research programs designed to define more accurately the risks from ionizing radiation. This does not mean taking more votes among the members of the standard-setting bodies and issuing more edicts, but rather conducting more epidemiological studies on human populations.

I would even consider lowering the level of radiation standards by a factor of ten *if* it could be shown that after a period of time all unnecessary exposure, especially medical, will be reduced and that there will be a net benefit to mankind by such action.

Protecting the Public

ALLEN BRODSKY

Allen Brodsky, a senior health physicist with the Nuclear Regulatory Commission, has helped develop radiation safety guidelines for universities, government laboratories, and hospitals. He is the editor of a series of volumes titled *Handbook of Radiation Measurement and Protection,* the first of which was published in 1979. Although Dr. Brodsky helped Dr. Thomas Mancuso establish the study of Washington's Hanford Atomic workers, he now disagrees with the findings of that study. In this essay Dr. Brodsky explains why he feels that current radiation standards provide adequate protection to both the public and atomic workers.

As a former radiation safety officer in several hospitals, universities, and laboratories, and as a professor at two universities, I have often explained radiation risks and methods of radiation protection to people of all levels of education. Therefore I know that the public can understand radiation and its relative risks, and participate in decisions on energy and the environment when the issues are explained properly in common language.

Making myself understandable is not enough, however. I also need to give you a basis for belief and trust. I will be satisfied only if I can help you to know the true nature of radiation and to learn to weigh the risk against the benefits. For ultimately, I believe, the welfare of society, the health and future of humanity, and the safety of my friends, family, and children depend on your knowledge of the subject.

More is known about radiation than is generally believed. Scientists have sufficient knowledge that if facts were presented properly, ordinary citizens could make judgments and vote on issues that could protect humanity and the environment, *without* overprotecting to the point of destruction. Undoubtedly you have heard conflicting reports. A few scientists (and some pseudo-scientists) who have views different from mine, and from the vast majority of scientists in fields related to radiation protection, have exaggerated the risks of radiation at every opportunity.[10] Let us call these scientists "antinuclear." They have underestimated our tremendous efforts toward radiation protection and the advances we have made. They have played up the most minor and insignificant of errors, as well as those that have been significant and that can and must be corrected.

Before refuting some recent claims, perhaps we should look for a moment at the nature of radiation. Radiation is a form of energy traveling from one place to another; it is a natural phenomenon. Radioactivity has always been in the human environment and in the human body. Consider potassium, an ordinary chemical found in the body which, like iron, is essential to good health. A tiny fraction of the potassium atoms, potassium-40 (K-40), is radioactive and one of the largest sources of natural radiation

Fig. 2–2 Drawing by S. Harris; © 1979 The New Yorker Magazine, Inc.

in the body. An average-size man ingests about 0.1 microcurie* of K-40 into his body, which exposes his tissues to about 20 millirems per year. This naturally occurring K-40 has always been a constituent of human tissue; man has not created this K-40.

In a year you receive more than twenty times the amount of radiation from the K-40 in your body than you do from routine emissions of radioactive gases from a nuclear power plant, which expose a neighboring population to less than 1 millirem per year.[11] You are even exposed to more radiation from cosmic rays in a high-altitude city like Denver, Colarado, or from several cross-country flights than from a nuclear power plant.

It has been argued that we have no choice in our exposure to K-40 or to the cosmic rays in the atmosphere, but that we can choose not to be exposed to a nuclear power plant's emissions. However, we can also choose not to live in a brick house, where the radioactive radon gas can expose a person to 150–300 millirems every year,[12] or choose not to have an annual chest x-ray, which exposes a patient to approximately 10 millirems, or choose not to build coal-fired plants, where the radon gas could expose the neighboring population to as much as 10 millirems per year (see Table 2–1). But before being alarmed by 1 millirem from a nuclear power plant, as compared to 6 millirems from a cross-country plane trip or 200 millirems from living one year in Denver, citizens should understand the nature of radiation and its effects.

Some critics of nuclear power have charged that a single ray of radiation can cause cancer. They do not explain that the risks of radiation effects are actually negligible at very low levels. In fact exposure even to a very high number of rays might produce no adverse health effects, so long as the "dose" is within legal limits. In one minute your body is bombarded from within by about 200,000–300,000 beta and many gamma rays from the K-40 mentioned earlier. If every ray produced harm we would be in big trouble. If we were very likely to contract cancer from exposure to radiation per se at these levels, diagnostic medical x-rays would hardly be worth the risk. The point is that the risk of cancer and genetic mutations from exposure to radiation results from *random* interaction between radiation and our cells. The risk can be made extremely *low* so long as individual doses are kept *low*.

A simplified model will help explain this random interaction. Gamma rays, like x-rays and other forms of radiation, are always emitted as discrete packets of energy, like bullets. Imagine that these bullets are being aimed by a nervous rifleman at two small targets several hundred yeards away. One of these targets is a golfball hanging from a tree by a flexible string. A baseball is hanging behind the golfball, blocking a hole in the tree. To give the tree "cancer," the golfball must be knocked into the hole in the trunk of the tree, where it will stay put once it gets in. However, the baseball is blocking the hole. Thus the first hit must break the string holding the baseball. Otherwise, even if the golfball is hit and heads toward the hole, it will hit

* A microcurie is a millionth of a curie.

TABLE 2–1   ANNUAL RADIATION EXPOSURE TO U.S. POPULATION

	Dose	Statistical Projection	
	Average (mrem)	Cancer	Genetic defect
SOURCE			
Natural Background	100	3050	193
Technologically enhanced[a]	5	150	10
Medical diagnostics	85	2600	164
Nuclear power (general public)[b]	0.03	1	0.05
Nuclear power (workers)	600	5	0.3
Nuclear weapons (development and fallout)	6	200	13
Consumer products	0.03	1	0.06
EFFECTS			
Total from all radiation sources		6007	381
Total from all causes, known and unknown		400,000	356,000
Percent from all radiation sources		1.5%	0.1%

Source: Adapted from the revised BEIR-III, "The Effects on Populations of Exposure to Low Levels of Ionizing Radiation: 1980" National Academy of Sciences, Washington, D.C., 1980.

[a] Primarily from naturally occurring radionuclides redistributed by human activities, such as mining and milling of phosphate and burning coal.

[b] Assuming normal operation, normal exposure, a population of 217 million people in the U.S., and that approximately 7000 person-rems would result in one fatal cancer.

the baseball and deflect. Eventually it will swing back to rest in its original position (analogous to biological "repair" of the initial damage). Thus two hits must occur and they must be made in the proper order. Neither hit would be likely for a nervous rifleman at such long range. It would be an unfortunate tree indeed that would contract cancer from two such rare hits in the right order, unless the number of bullets spent began to reach some enormous number. Even so, for two or more bullets one can easily see that the chances of "tree cancer" are not zero, and that they increase with increasing exposure to bullets. The chance nature of radiation effects in humans at low-dose levels is similar to the bullet-golfball-baseball analogy just described.

Critics of nuclear power have charged that a look at the history of radiation standards, where levels of exposure to the public have been reduced by a factor of five since 1934, serves as proof that we have consistently underestimated the dangers of radiation. How can we be sure, they ask, that we have arrived today at an adequate standard for protection, in light of inadequate former standards?

These concerns may seem valid, unless you familiarize yourself with the true basis for this lowering of dose limits.[13] From 1934 to 1950, after data on the effects of radiation were accumulated from hundreds of animal, plant, and epidemiological studies, the maximum permissible exposure for workers was lowered from 0.1 R per day to 5 rems per year. Along with these limits, the standard-setting bodies have also recommended keeping exposures to a minimum. As a result of this philosophy of ALARA, the vast majority of radiation workers receive below one-tenth the permissible limits. Mounting evidence presented to the standard-setting bodies since 1956 confirms the belief that current recommendations of 5 rems (5000 millirems) per year as a limit for workers in the nuclear industry, and 0.5 rem (500 millirems) per year for the public, have not caused substantial harm.[14]

In the past decades hundreds of epidemiological studies have reached consistent conclusions regarding the effects of radiation. In addition to the Hiroshima-Nagasaki data and the studies of British x-ray patients treated for arthritis of the spine, consider just a few findings from many further studies:[15]

○   Thousands of American, Czechoslovakian, Swedish, and Canadian uranium miners have been exposed to radon gas and decay products. One follow-up study from 1950 to 1963 showed that 22 of the 3415 workers exposed developed lung cancer. Since five cancers would normally be expected from a similar, but unexposed, population, we could calculate that there were 17 excess cancers among the miners. (These studies also showed the possibility that smoking was at least in part the cause of these excess cancers.)

○   One study of 3521 radiologists showed that 438 deaths resulted from leukemia and other cancers, whereas 295 deaths would be expected from an unexposed population. (This study was conducted in the early years of radiology when the doctors might have received thousands of rems over a lifetime. Those doses are far greater than doctors or radiation workers have received for the past twenty or more years under current regulations.[16])

○   Between 1926 and 1957 many infants less than six months old received anywhere from 61 to 600 rems from chest x-rays to treat an enlarged thymus. Of the 2207 infants studied, 13 developed leukemia and thyroid tumors, 6 of which were benign. Since two cancers would be expected on the basis of the control population, there were said to be 11 excess cancers.

○   27,793 women with cancer of the cervix who were treated with x-rays (300 to 1500 rads* to the bone marrow) actually developed *fewer* cancers than would be expected from a normal population: 10 developed leukemia and lymphatic malignancies, whereas 11 cancers would be expected from an unexposed population.

Populations as diverse and unrelated as uranium miners, Japanese casualties at Hiroshima and Nagasaki, and children exposed to x-rays show remarkably similar cancer rates. For example, if we exposed 100 million people to 1 millirem each, an in-depth analysis of the uranium miner data would show that we could expect 0.05

---

* Whereas the rem is a measure of biological damage, the rad is a measure of absorbed energy. For gamma and beta rays, exposure to 1 rem is approximately equal to exposure to 1 rad. For alpha rays, 1 rad = approximately 10–20 rems. By definition, a rad is 100 ergs of energy absorbed by 1 gram.

lung cancers per year. Similarly the Hiroshima-Nagasaki data would show 0.06 lung cancers per year. In other words the data are extremely consistent, and the number of expected deaths has been very small. Of the tens of thousands of patients studied over the past several decades, only a handful have developed cancers. Hundreds of these studies have been reviewed by the standard-setting bodies, who would have been compelled to lower the recommended limits had the scientific evidence indicated that radiation was more harmful than previously believed.

Some critics have challenged the reliability of these epidemiological health studies, stating that external factors unrelated to radiation may have biased the results. It has been cited that the Hiroshima-Nagasaki casualties were also exposed to the trauma of the bombings, and that only the stronger Japanese lived long enough to contract cancer. However, no *quantitative* evidence is offered to support this thesis. Without concrete data such critics may well be imagining that there is a correlation between surviving the aftermath of the bombings and being slightly more resistant to cancer. Unless you can separate these factors scientifically, it is impossible to draw conclusions. In the absence of solid quantitative evidence it is easy to hypothesize about an entire series of speculative biases. At any rate the consistency between the Hiroshima-Nagasaki data for leukemia and data from other studies still indicates that the risk estimates of organizations of experts are the ones to rely on.

Furthermore, scientists never rely on simply *one* study. Since a "perfect study" does not exist, it may well be true that a few studies have biases that we do not recognize. However, hundreds of our studies are in agreement, within a factor of approximately two or three. The *consistency* between diverse studies conducted on an international level and the repeated similarity of our results have convinced the vast majority of the world's leading scientists to accept the results.

In other words, there are comfortable margins of safety built into these radiation standards. For example, in 1981 scientists at the Lawrence Livermore National Laboratory discovered that radiation from the Hiroshima bomb was previously miscalculated and careful reanalysis of the revised data showed that there may be some minor changes in the risk factors for certain forms of cancer. However, overall these risks are still well within the margins of safety originally included in our radiation standards. Even if the risk for certain cancers goes up by 100 percent, no changes in our radiation standards would probably be required because radiation standards have been extremely conservative to begin with. Dr. T. Straume, one of the scientists at the Lawrence Livermore National Laboratory who was involved with reevaluating the Hiroshima data, stated in June 1981[17] that these revisions do *not* effectively change the risk coefficients found in the National Academy of Sciences' report on the Biological Effects of Ionizing Radiation (BEIR).[18]

Therefore I feel confident in stating that exposures to 1 millirem would be expected to produce no more than 13 chances in 100 million persons of a future cancer death (see Table 2-2).[19] This means that the risk to people in the vicinity of the Three Mile Island accident at Harrisburg, Pennsylvania, who received an average dose of 1

TABLE 2–2 CANCER RISK FROM RADIATION

Type of Cancer	Maximum Chances of Cancer Death per Rem	Average Shortening of Life
Leukemia	23 in a million	0.20 days
Lung	25 in a million	0.14 days
Stomach	12 in a million	0.065 days
Alimentary canal	4 in a million	0.0073 days
Pancreas	4 in a million	0.0073 days
Breast	29 in a million	0.062 days
Bone	5 in a million	0.029 days
Thyroid	10 in a million	0.0062 days
All other	20 in a million	0.11 days
Total[a]	13 in 100,000	0.63 days

Source: From R. L. Gotchy, "Estimation of Life Shortening Resulting from Radiogenic Cancer per Rem of Absorbed Dose," *Health Physics* 35 (1978): 563–65. Revised to give the maximum chances of cancer and life shortening to be expected by an individual worker.

[a] Studies are conducted to determine the cancer risk from radiation exposure to each organ of the body. By adding these risks, scientists arrive at the cumulative risk to the entire body (i.e., for each rem of exposure we have 13 chances in 100,000 of dying of cancer, or for 1 millirem of exposure, 13 chances in 100 million).

millirem[20] would be no greater than 13 chances in 100 million of dying of cancer.[21] Compare this risk to the 1 in 100 chance of being injured or killed in an auto accident during your lifetime. Certainly the scare stories about the radiation effects of Three Mile Island would be discounted if people realized that the risks from that exposure were minimal.[22]

The Benefits of Radiation

I have talked at length about the ways in which the dangers of radiation have been distorted. Now let us turn to the many major benefits to be derived from radiation, aside from the electricity generated by nuclear power plants.[23]

Radiation has cured hundreds of thousands of patients and improved the quality of our lives. When we ride safely and comfortably in high-speed trains, jet planes, or cars, we have the satisfaction of knowing that the machinery has been tested for strength and reliability by radiographs made by high-energy gamma rays from by-products of nuclear energy. Gamma rays and x-rays from radioactive materials are used to detect the smallest defects in turbine blades, high-pressure steam boilers, and other potentially dangerous machinery. Mechanical failures and transportation accidents have been greatly reduced, while the comfort and safety of the ride have been improved.

In medicine, radiation and radioactive pharmaceuticals are among the chief diagnostic tools in every hospital in the country. We tend to forget that before the use of radioactive materials in nuclear medicine, doctors often relied on exploratory surgery or educated guesses to diagnose patients. Today doctors routinely inject minute quantities of radioactive chemicals into our bodies to examine the health of our organs, locate tumors, detect malfunctions of the thyroid glands, and determine the chemical balances in our bodies. Radiation is also used in cobalt therapy to treat *one-half* of all cancer patients. (Radiation can cure certain forms of cancer by killing cancer cells much more readily than healthy cells.)

In farming and agriculture, radiation has helped keep down our food prices by ensuring the health of farm animals and crops. For example, at one time the screw worm fly was infesting cattle throughout the Southwest, which resulted in higher beef prices. By using radiation to sterilize the screw worm fly, this multi-million-dollar pest was eradicated almost overnight.

In criminology, radiation has opened up a new area of crime detection. Tiny fragments of paint, hair, cloth, dirt and so forth left at the scene of a crime can readily be identified by radiation. Scientists irradiate the fragments with neutrons until the samples themselves become radioactive, whereupon they can be identified much as if each sample were fingerprinted. Using this same method of detection, police have exposed art forgeries by having a laboratory examine the chemicals found in the paints.

Thus radiation is used in myriad diverse ways in our modern society. It is used to measure anything from a wine's copper content, which has an important bearing on the taste, to the chemical content in a river, where a nearby factory could be discharging poisonous heavy metals into the water.

The Mancuso Study

Despite the substantial benefits of radiation and evidence in favor of current radiation standards, people fear that current limits may be inadequate. Why?

Widely publicized recent studies, most particularly Dr. Thomas Mancuso's study of radiation workers at the Hanford Atomic Works in the state of Washington,[24] claim that occupational radiation risks are higher than the government's recommendations acknowledge. A number of scientists (who I believe are eminently qualified to analyze epidemiological data[25]) and I reject Dr. Mancuso's claim that a significant or even detectable number of cancers has resulted at Hanford from occupational exposure to low-level radiation. My opinions are based on my personal knowledge of the study: I worked with Mancuso and his co-worker, Dr. Barkev S. Sanders, for the first eight years of the study, and participated in presenting and publishing the preliminary results.*[26] I later read the claimed results as they were presented by Dr. Alice Stewart in 1976 and published in 1977, four years after I left the study.

*From 1964–1972 while I as on the faculty of the Graduate School of Public Health, University of Pittsburgh, Dr. Mancuso employed me as a co-investigator to help collect and interpret radiation exposure data and establish proper methods of conducting a radiation epidemiological study.

The methods Dr. Mancuso used in drawing his conclusions bore no resemblance to the methods developed initially by Dr. Mancuso, Dr. Sanders, and me to carry out a thorough, long-term study of mortality. From among the numerous errors and obvious faults of the Stewart presentation, only a few of the logical errors are mentioned below as examples.[27]

First, in a proper scientific study the established method is to compare the cancer rate in the irradiated population with the cancer rate in a nonirradiated control population. No proper control population was included in Mancuso's final analysis, although such populations were constructed originally by Dr. Mancuso and Dr. Sanders as an essential part of the project.

Second, the data have been reanalyzed by two independent groups[28] and the only excess cancers they found were multiple myeloma (a type of bone marrow cancer) and cancer of the pancreas. There is little evidence of excess cancers of these types among the Hiroshima-Nagasaki casualties, whose number was ten times more numerous and who were exposed to larger radiation doses. It seems reasonable to assume that the four extra cases could have been caused by some type of chemical or other exposure, a possibility which Mancuso had originally insisted on investigating when Dr. Sanders and I conducted the project, but which he later ignored when he brought in Drs. Alice Stewart and George Kneale. The Manusco analysis ignored outside factors that could very well nullify the results. The workers' higher than expected exposure might be explained by radiation exposure at installations other than Hanford, ordinary medical exposures, and chemical exposures. In fact Mancuso failed to go back into the plant records—as he always insisted must be done when I was with the project—to see if workers had been exposed to chemicals or other carcinogens.

Third, the Mancuso analysis neglected to consider obvious reasons, totally unrelated to radiation, that could explain why those who died of cancer had slightly higher radiation exposures. For example, the higher exposures could be due to the fact that people who die of cancer are generally older, and therefore have a longer work history with more exposure to radiation. But this should not necessarily imply that radiation *caused* these cancers, as Mancuso and co-workers claim.

In prior studies Dr. Mancuso made important contributions in alerting society to the hazards of exposure to certain chemicals. Similarly, Dr. Stewart alerted society to the hazards of x-rays received by fetuses *in utero*. It is ironic that I must now point to the absurdity of the Mancuso study as analyzed by Drs. Stewart and Kneale. However, a dangerous precedent is being set in our democratic society if claims of important health findings can be given credence on the basis of such distortions of scientific method. At best, *if* the analysis had been carried out properly, the Mancuso-Stewart-Kneale approach to evaluating the study data would have been used to suggest a trend for radiation effects. Yet these results were put forth as strong conclusions of *proven* cancer causation.

There are those who claim that the Atomic Energy Commission discontinued support of Mancuso because the AEC wanted to cover up his "positive findings" that exposure to radiation at Hanford was causing cancer. I was present at the meeting in 1964 when the AEC initially asked Mancuso to conduct the study. Does it make sense for the AEC to have asked Mancuso, who had *already* found chemical causes of cancer in industry, to conduct the study if the AEC were trying to engineer a coverup? The AEC provided Mancuso with more than $6 million in support over almost a dozen years before refusing to renew his funding. The AEC took this action because Mancuso refused to publish his preliminary results, or even his planned methods in sufficient detail. This was a year and a half before he presented any so-called positive findings.[29]

The media further distorted the truth by widely publicizing Mancuso's results but failing to mention that many competent scientists, whose studies unfortunately were less publicized, faulted his methods. At a congressional hearing in 1978 when I refuted Mancuso's claims before the House Subcommittee on Health and Environment, I was appalled to observe that the media apparently did not take my remarks seriously. I noticed that the television camera lights were turned on for Mancuso's testimony and then turned off while I (and other reputable scientists who opposed Mancuso) testified.

Ensured Radiation Protection

Health officials like myself constantly review our standards and any new epidemiological studies that appear in order to maintain the best radiation protection that is possible. However, further improvements are needed in government support, as well as follow-up studies of both workers and members of the public exposed to radiation. The present divisional organizations within the Nuclear Regulatory Commission (NRC) are not adequate to deal with the important and complex health questions now confronting the NRC. Present divisions have too much to handle: questions of siting alone draw on the expertise of geologists, seismologists, hydrologists, meteorologists, environmental scientists, economists, health physicists, radiation biologists, and radiation dosimetrists. Two new divisions, managed and staffed with top health science professionals, should be established within the NRC: divisions of *Health Science Research* and *Health Data and Risk Evaluation*. Presently we evaluate numerous claims, calculations, and epidemiological studies. This is an important task, but no substitute for our own program of carrying out, as well as monitoring, health follow-up studies. These changes would more likely involve less cost to the taxpayer, rather than more, as a result of more efficient management and coordination of efforts already underway.

When the NRC was born out of a portion of the AEC little competence in broad areas of the health sciences was brought along. Although many improvements have been made, management has been dominated by nuclear engineers to the point where the NRC has been almost a closed fraternity to top health scientists. The time has come for a new and proper balance—one which includes health professionals who have earned reputations for managing research programs in radiobiology and radiation risk evaluation.

There are critics outside (as well as inside) the NRC who believe that health studies conducted within the NRC would not be objective. This is a preposterous, defeatist attitude. One important way to ensure objectivity and verifiability would be to establish an Advisory Committee on Radiation Health Effects, paralleling our Advisory Committee on Reactor Safeguards. Comprised of independent leading health scientists, the committee would help assess all relevant health risk data and radiation exposure standards for the commission, and serve as an oversight committee to ensure objective in-house health follow-up studies. This committee would provide a trusted and independent group of experts who also could advise Congress directly. If we assume that the NRC's health studies cannot be made believable to the American public, when NRC staff members know the industry and data best, then how can we justify the very existence of the NRC?

Within the past thirty years the most prestigious scientific bodies in the world (the ICRP, NCRP, and UNSCEAR) have formulated radiation protection standards. Government regulations to ensure good radiation safety practice have followed the standard-setting bodies' recommendations. Over three decades, because of the effectiveness of government controls, the average worker monitored for radiation exposure has received less than 5 percent of the maximum permissible limits. Although additional studies might be undertaken or minor adjustments made to our standards and procedures for the purpose of updating our information, current permissible limits are quite adequate where radiation exposure is concerned.[30]

I hope that we will be able to make further improvements in reactor and facility design and in planning and managing radiation safety programs. However, the degrees of risk currently presented by the nuclear power industry and its by-products are no cause for the alarm that has been generated by the antinuclear movement.[31] With proper health and safety management, a nuclear power industry can and must operate safely in our nation. We need nuclear power to provide the essential energy for a healthy society, gradually replacing the more polluting and nonrenewable fuels such as coal and oil, which should be used to make valuable fertilizers and petrochemicals for further generations.

George Orwell Understated the Case

JOHN W. GOFMAN

In the early 1940s John W. Gofman co-discovered uranium-233 with Glenn Seaborg, and went on to join the Department of Medical Physics at the University of California at Berkeley in 1947. From 1963 to 1969 Dr. Gofman was associate director of the Lawrence Livermore

Laboratory in California, one of the nation's foremost weapons laboratories, where he broke ranks and stated that thousands would die if the maximum allowed limit of 170 millirems of radiation per year were reached. He is now chairman of the San Francisco–based Committee for Nuclear Responsibility, professor emeritus of medical physics at the University of California at Berkeley, and a member of the clinical faculty of medicine at the University of California at San Francisco. Dr. Gofman here denounces the pro-nuclear scientific community, charging that their systematic policy of "disinformation" has deceived the American people, and may result in thousands of needless cancer deaths.

When experts disagree, whom shall we believe? Certainly everyone has asked this question, or heard it asked, again and again and again. One of the hopes for this book has been that if both sides of the question of nuclear power are presented fairly, the public will then be able to make up its mind. But perhaps this solution does not come to the heart of the matter. If the issue of nuclear power is ever to be resolved, it will be necessary to consider: (a) the identification of individual human rights; (b) the proper role of government; (c) the role of information and disinformation in science; and (d) the source of funding for science research and scientists.

The Identification of Individual Human Rights

Who can doubt that respect for the individual human rights of others is paramount? Each of us must respect the rights of all other individuals to live their lives as they choose so long as *they* respect the rights of all other individuals to have precisely the same rights. This means that each of us has the right to life, to liberty, and to property, so long as we do not interfere with the equal rights of all others. There is *zero* room for any coercion of one person by any other person, group, association, or organization, be it a corporate organization or government. Corporations, associations, and governments acquire no rights beyond those which inhere to the *individuals* who make up such groups. The *only* proper function of government is to ensure that no individuals or groups violate the natural, inherent rights of others, whether those violators are within or outside the country in which we live.

Unfortunately this concept of liberty, which is what I think was intended for the United States, has become sorely distorted. While we may agree broadly that the individual human rights described above are paramount, it is inevitable that some individuals and groups of individuals decide not to accord others their rights. Government, supposedly set up for the protection of individual rights, has become a coercive force which accords certain individuals, groups, and corporations special privileges in the form of regulations, laws, grants, and freedom from liability. We all know that the government handsomely funds the research and development of the nuclear industry, a role that is totally inappropriate for government. We also know that the government frees the nuclear industry of liability for the life and property destruction it can potentially produce in its grand experiment in containment of radio-active poisons. Many people have been taken in by the idea, widely promulgated, that these additional functions of government are part of the fabric of *social good*.

Pollution and Trespass

Energy corporations deserve *no* special privileges. They have only the same rights as does any individual in the society, and they must respect every feature of the rights of all individuals in the society.

In the earlier years of our society the idea that any company, corporation, or individual could trespass upon the property (including the body) of an individual was not accepted. Sadly, that has changed. In the course of the Industrial Revolution the courts have prostituted the idea of individual rights against trespass for the "social good." Indeed, everytime individual rights are violated by government, be it the legislative, executive, or judicial branch, it is always held to be for the social good. But there is no social good which justifies violations of individual human rights. And so it was a fraud against human rights when the courts began to hold that companies could pollute the air, the streams, and people's bodies.

Some persons and groups are willing to concede that it would be wrong to trespass upon the body of a person with pollutants if the person can show he is being harmed by the pollutants. Whether or not pollution is harmful is irrelevant; every individual has the right not to be polluted because it violates our freedom from trespass upon our bodies and property. In fact requiring proof of harm is dangerous, because it can ultimately lead to genocide frrom the aggregate late effects of many small risks from pollutants. If corporations had to recognize that pollution is a major crime, techno-logical-industrial activities would be conducted in a totally different manner, which, incidentally, would not lead to any loss of the fruits of technology and industrializa-tion.

It is cheaper to pollute than it is to control pollution at the source. Polluters are willing to risk the lives of others *provided* the company or group violating the lives of others is protected by government from financial liability for such actions. This is virtually industrial piracy and the opposite of a true free market, where individual rights receive total respect and where financial responsibility is shouldered by those who create risks.

The Benefit-Risk Analysis

One of the cleverest, most diabolical techniques worked out to violate individual human rights by pollution is the *benefit-risk* doctrine. This doctrine holds that there is a social good which must take precedence over individual rights. Thus government intercedes on behalf of an industrial technology not only by stealing taxpayers' money to fund the research and development that industries should do for themselves, but also by evaluating the "benefits" of certain industrial technologies and the "risks" of these same technologies, to determine appropriate "standards" of pollution which shall be acceptable to the public in exchange for the "benefits" the public will reap. This coercive agency, government, makes these decisions for us, and we must obey them for the social good. Once one concedes that the benefit-risk doctrine has any validity at all, one has conceded the whole ball of wax, since there are no limits to the application of the "benefit-risk" doctrine.

If the modern-day "environmental impact statement" were in vogue in Hitler's Germany, there would assuredly have been benefit analyses to prove that genocide of six million Jews was for the "social good." There may be those who read this and say I have gone too far. Every word is meant as stated here.

Health Effects of Radiation and the
Dangers of the Benefit-Risk Doctrine

Since the question of "which experts to believe" can be asked in the field of health effects of radiation, we should examine the folly of the benefit-risk doctrine to see why controversy even exists at all. With government and industry working hand in glove to promote nuclear power, it is fair to say that government, on behalf of its industrial friends, has a vested interest in making nuclear power acceptable to the populace. And since the benefit-risk doctrine is the standard mechanism for making polluting technologies acceptable, the problem is simply one of assuring that the benefits appear to be much larger than the risks. And since this benefit-risk analysis has to be marketed to an unsuspecting public, it must be made to appear objective, authoritative, and free of special interest.

This is where scientists and academia enter the picture. Their role is to provide the mantle of credibility for the information. Government is happy to acquire all the information which will support the benefit side of its analysis and minimize the risk. But if the *information* does not support a program to which it and its special interest friends are committed, then the government is equally interested in acquiring all the *disinformation* possible to support its case, and the services too of those who will help spread both the favorable information plus the requisite disinformation.

Lest the reader misunderstand, I am not suggesting that scientists are asked to lie. Indeed the fabrication and spreading of disinformation is far more sophisticated than that. The half-truth or the meaningless statement are good examples of disinformation, particularly when spread by a professor of high repute. A professor of medicine might say, "I am not aware of a single case in my practice of a lung cancer caused by plutonium." As long as no one asks him how he would know one when it stood before his nose, he has successfully provided his contribution to the desired disinfor-

Fig. 2–3 **"First we have to convince the people that good health isn't everything."** [© 1982 S. Harris.]

mation. He can even be incensed if someone accuses him of lying, for *he has not lied*.

Another common statement is "No studies of humans have shown that low-dose radiation causes disease XYZ." If no one asks the simple question "Have *any* studies been done?" the statement goes unchallenged and succeeds in spreading its message of disinformation. The man on the street goes away contented to know that no study has shown radiation causes disease XYZ. Marvelous!

One of the endearing traits of the public is its willingness to believe that it will not be deceived, particularly by government, although that willingness has been eroded to an extreme degree recently for excellent reasons. But even today one often hears, "If that were really dangerous, the government would not allow it." Such trusting is a fine trait, and it is sad that such crass advantage is taken of this willingness to give people the benefit of the doubt.

A look at the history of the Atomic Energy Commission/Energy Research and Development Administration/Department of Energy (AEC-ERDA-DOE) reveals classic cases in the field of generating disinformation. AEC-ERDA-DOE has expended huge sums of public taxpayer-funded grants to national laboratories and to universities to assess the health effects of ionizing radiation. It is essential to be fair and to state that some very good research has resulted from such funding and has provided *information* which is quite useful. But the scientists come to know what information is desirable and what is definitely not desirable.*

I was a director of one of the twenty-odd biomedical laboratories sponsored by AEC. The directors of laboratories used to get together every three months or so to discuss findings, and at one of those meetings I brought up some considerations of deaths to be caused by ionizing radiation in humans. One of the directors of a major national laboratory biology and medicine department took me aside in the hallway after that session and informed me, "You have just committed a very serious error." What error? I was then told that mention of deaths from radiation was simply not one of the subjects that the AEC liked to hear about. "Just don't talk about it," I was advised.

Trouble in Paradise: The Phase of Committees and Commissions

Every once in a while in the paradise of information-disinformation of such agencies as AEC-ERDA-DOE, a dissenter or two comes along and this causes much temporary inconvenience. The reason is that by the time they are silenced or left without funds to continue, the issue may have received a fair amount of public attention, and then we must enter a new phase—the phase of committees and commissions.

In 1970 Dr. Arthur Tamplin and I published estimates[32] that the then-permissible exposure limits, *if reached,* would result in 16,000–32,000 extra cancer deaths per year in the U.S. This was much higher than any prior pessimistic estimate. As a result of the pressure of comments by Ralph Nader and Sen. Mike Gravel (Alaska), Secretary of HEW Robert Finch announced that he was asking the National Academy of Sciences to appoint a committee to investigate our findings. That was the birth of the BEIR (Biological Effects of Ionizing Radiation) Committee. Although the committee was set up in direct response to our work, no member of the committee has contacted me in its existence.

The BEIR Committee published its first report, BEIR-I, in 1972,[33] suggesting that Tamplin and Gofman had overestimated the number of cancer deaths to be caused by ionizing radiation by a factor of about five. Considering that the AEC was previously unwilling to concede *any* deaths from the "permissible" doses of radiation,

---

*For example, after decades of financial support of scientists it was rare indeed to find scientists from those handsomely funded university and national laboratories publishing calculations of the numbers of cancer deaths to be expected from radiation at the "permissible" doses. It was Dr. Linus Pauling, who was not AEC-supported, who began to calculate the number of leukemia and cancer deaths for various doses of radiation, and who *told* the public of the results. AEC-supported scientists may well have made similar calculations, but *their* contribution to disinformation lay in the failure to notify the public of their findings. It would seem almost impossible statistically that no AEC-supported scientist found Pauling's calculations reasonable. The answer was very clear: the AEC was buying the services of scientists and professors to "study the question" but not to find answers which were unwelcome for AEC's promotions.

agreement within a factor of five was taken by the scientific community at large as confirmation of the work of Gofman and Tamplin. By the time the committee had reported its findings, we were quite sure that we had still *under*estimated, not over-estimated the cancer price of radiation exposure. So it was of the greatest interest to learn how the BEIR Committee had managed to estimate that Gofman and Tamplin were too *high* by a factor of five.

We used the "relative risk" method of calculating the excess cancers caused by radiation, which I shall explain shortly. One major source of epidemiological data available at that time was the series of British patients with an arthritis of the spine who had received x-ray therapy. Crucial evidence rested on the radiation-induced stomach cancer in those patients. In our analysis we used a value of 61.6 rems as the dose to the stomach in these patients, a value estimated by Drs. G. W. Dolphin and I. S. Eve from the United Kingdom Atomic Energy Authority in Harwell, England.[34]

The BEIR Committee elected to center its critique of Gofman and Tamplin's work on this dose of 61.6 rems to the stomach. If the committee could show that it took more than 61.6 rems (a higher dose of radiation) to induce the same number of cancers, then the harmful effects of radiation would be reduced. The BEIR Commit-tee reported: "It is clear from the above that the estimation of the mean stomach dose from the mean dose to the spinal marrow is crucial. The committee has reconsidered this question and has concluded that the previous estimate by Dolphin and Eve is seriously in error. A full discussion is given in Appendix V. It is sufficient to state here the conclusion, namely, that the minimum value of the dose to the stomach is 250 rads, and a value of double this is not unlikely."[35]

The BEIR Committee, in an elaborate appendix calculating stomach dose, *but signed by no one, ostensibly* had removed the basis for the use of 61.6 rems and replaced it by a value four to eight times higher. The effect, of course, was to indicate that those parts of the Gofman-Tamplin calculations which relied on the stomach dose must be too high (in radiation risk) by a factor of four- to eightfold. The BEIR Committee does not *suggest* the Dolphin-Eve dosimetry was in error; instead it asserts positively that the dosimetry was wrong by a *minimum* of fourfold and possibly by eightfold.

The truth ultimately emerges in science, however, despite the best of efforts to becloud it. Seven short years later, in the preliminary BEIR-III report of 1979, there is a table in which this same stomach dose to the arthritic patients is presented as part of a reanalysis of dose from x-rays being done in Britain.[36] Amazingly enough, in the new table the dose to the stomach is presented as between 67 and 89 rems for those patients. The average of these values is 78 rems, exceedingly close to the 61.6-rem value used (from Dolphin-Eve) by Gofman and Tamplin. So the profound and assertive calculations of the BEIR-I Committee, which are now acknowledged to be in error by a factor of between 3.2 and 6.4, and which were used to prove that Gofman and Tamplin had overestimated the cancer risk of radiation, *simply no longer exist*. This was not a minor blunder which the BEIR-I Committee had made, but rather a very serious and major error of substance. Yet, at total variance with scientific custom and propriety, both the Beir-III Draft and Final Reports are strangely *silent* about its large blunder in the dosimetry, and strangely silent about its previous erroneous criticisms of the calculations of Gofman and Tamplin.

Absolute Risk Versus Relative Risk

Once Pandora's box had been opened and antinuclear activists knew about the existence of calculations of the actual human cancer deaths to be caused by ionizing radiation, quasi-official and governmental bodies announced their "scientific preference" for the "absolute risk" method of calculation of cancer deaths from radiation. The absolute risk method conveniently gives low estimates for the risk.)

The "absolute" method of calculating cancer risk per rem simply counts the number of extra cancer fatalities at some period after irradiation, divides that number by the radiation dose, and then presents a value for extra cancers per rem per year for, say, a million exposed persons for some number of years beyond the exposure. Since 1969 Dr. Tamplin and I have repeatedly pointed out that this procedure will result in an erroneously low estimate of the cancer risk of radiation, simply because the absolute value for the excess radiation-induced cancers in the early years of a follow-up of exposed persons is much lower than it will be when the full effect is manifested. While governmental bodies consistently announce their "scientific preference" for the absolute risk method, scientific *evidence* in favor of their preference is notable only by its absence. Grudgingly, such committees as the BEIR and the U.N. Scientific Committee on the Effects of Atomic Radiation (UNSCEAR)[37] do present the 'relative risk" estimates, but with sufficient caveats of a negative nature as to discourage use of the relative risk method. Let us now examine the scientific evidence itself, which proves the error of the absolute risk method.

When a group of persons has been irradiated, at a specific age, and then followed for the development of cancer, two values are obtained: the *observed* value, or O value, which is the number of cancers in the irradiated group, and the *expected* value, or E value, which is the number of cancers in a group otherwise comparable but unirradiated. All the human epidemiological evidence available concerning the ratio O/E as a function of time after irradiation shows that the O/E value is increasing with time after irradiation at least out to thirty years beyond irradiation, and more likely out to beyond forty years beyond irradiation.[38] I have analyzed in detail the data on this issue from five studies.[39]

There can be no doubt, after examination of these human epidemiological data, that the O/E values are rising for at least three decades after irradiation and possibly rising for a period longer than that. But what is of even much greater importance is that in *all* of these studies, this rise in O/E values is occurring at the same time that E itself is rising, and in most cases rising rather steeply. A little arithmetic would

TABLE 2–3   PEAK PERCENT EXCESS CANCER RISK PER REM
FOR IRRADIATION AT VARIOUS AGES

Age at Irradiation	Peak Percent Increase in Cancer Fatality Rate per Rem of Whole-Body Exposure
0	81.4
6.9	54.8
14.5	4.4
25.9	3.7
31.2	3.8
42.3	0.9
50+	0.04

show that if O/E is rising, and if E is *also* rising, then the *absolute* number of radiation-induced cancers per year is still rising for two reasons some three decades beyon the irradiation. Since the vast majority of follow-up studies are for times short, o the average, compared with three or four decades, it is clear that a *gross underest mate* of the cancer risk will occur if the absolute method is used to calculate th cancer price of radiation exposure.

On rational, scientific grounds there is no way to make the absolute method o radiation risk assessment consistent with reality. But is rational science the issue?

Estimating and Underestimating the Cancer Risk of Radiation

I prefer to describe the cancer risk of radiation in terms of a concept known as th *cancer-dose*. The *cancer-dose* is defined as that number of person-rems introduce into a population sample which will guarantee one fatal cancer in the remainin lifetime of the exposed population sample. I have described the method for calcula ing the cancer-dose in several places, with updating as new epidemiological da became available.[40] In my most recent estimates I have presented the reasons fc using a forty-year point as the peak for O/E values for solid cancers (excluding lei kemia and bone sarcoma), which also means that the excess percent cancer risk pe rem is reached in the fortieth postirradiation year. I have calculated the cancer dose from those peak percent values.[41] By using all the available human epidemiologic studies of low-LET radiation* which qualify by having meaningful dosimetry an meaningful follow-up periods, the *mean* values in Table 2–3 were arrived at for th peak percent excess cancer risk per rem, for irradiation at various ages.[42] This gel

---

*Low-LET radiation is low-energy Linear Energy Transfer radiation, which includes beta and gamma ray Alpha rays, which transfer more energy per inch than beta or gamma rays, are called high-LET radiatior

:ral type of distribution of mean values, with enormously greater sensitivity shown 'or irradiation in the early years of life, is consistent with the same type of distribution shown originally, with a fixed percent increase per rem beyond the latent period, by Gofman[43] and in the BEIR-I Report (1972). In spite of the overwhelming evidence 'or the greater sensitivity of those young at irradiation, government and quasi-official bodies persist in making erroneous statements which indicate they refuse to acknowledge the enormous sensitivity of the young to the development of radiation-induced cancer.[44]

If one wonders why the governmental bodies are so refractory to admission of the very great sensitivity of the young to cancer induction by radiation, the answer is obvious. If a population of mixed ages is irradiated, those under ten years of age contribute on the order of half of all the cancers induced.[45]

Three Mile Island

The Three Mile Island accident provides an excellent illustration of the proper use of the cancer-dose in estimating the consequences of population exposure to whole-body radiation. For a stabilized population of mixed ages I have presented the following cancer-doses:

> For males (U.S.) 235 person-rems of whole-body radiation is one cancer-dose.
> For females (U.S.) 300 person-rems of whole-body radiation is one cancer-dose.[46]

Since our population is very roughly half males and half females, it is acceptable to use the average value, 268 person-rems, as the cancer-dose for the mixed population, overall.

Since the dosimeters to measure the radiation were so sparsely located in the surrounding area, we shall never know the true dose delivered to the population in the environs of the Three Mile Island plant. But some measurements were available. Usually an episode of radiation release is handled with the familiar public relations statement "there will be no harm to the public health." Even though that was tried by Metropolitan Edison and the NRC early in the accident, TMI was simply too big an affair for standard disinformation.

Our cancer-dose for such a population is really 268 person-rems, not the 10,000 person-rems estimated erroneously by UNSCEAR.[47] So if we believe the government's estimate of 5000 person-rems as an upper limit to dose, then *19 fatal cancers* (5000/268) were guaranteed by the accident. Even though the monitoring was grossly inadequate, the governmental sources indicate that 50,000 person-rems is out of the question as the dose received. Since no one really knows, I do not think we can discredit this as a possible dose. If the dose were truly 50,000 person-rems, the number of fatal cancers guaranteed would be 187, which would deserve a major disinformation campaign.

The Linear Hypothesis

For years governmental scientists and committees accepted the linear hypothesis as the basis for calculating the cancer effects of radiation, as the "prudent" thing to do. It was "prudent" until significant radiation releases began to occur. Then the acceptance of this hypothesis became a public relations albatross. Soon a rash of statements began to appear, with prominent scientific names in authorship, that the linear hypothesis grossly *overestimates* the cancer risk of radiation.

What surprises me is that so many scientists are now making scientific fools of themselves, because the evidence is steadily accumulating in powerful fashion to show that the linear hypothesis probably underestimates the true cancer risk of radiation at low doses. The scientists are coming out with the wrong piece of disinformation at the wrong time.

I have done an exhaustive analysis of the worldwide epidemiological evidence by asking a very simple question, scientifically: "If we measure the cancer rate induced *per rem* for high radiation doses and contrast that measure with the cancer rate induced *per rem* for low radiation doses, *which* is higher, if they are not identical?"[48]

The linear hypothesis would lead one to expect the *same* cancer risk per rem at high and low total doses. The favorite promotional hypothesis is that known as "diminishing effect per rem at low dose." This suggests cancer risk to be related to a power of dose higher than one. There is a third possibility, *supralinearity*, with the cancer induction effect being even higher per rem at low total doses than at high total doses. This hypothesis suggests cancer risk to be related to a power of dose less than one.

Four groups, with large numbers of cases for analysis, were studied, including the following: (1) overall cancer death rates in Hiroshima and Nagasaki; (2) leukemia death rates in Hiroshima and Nagasaki; (3) breast cancer incidence rates in Hiroshima and Nagasaki; (4) breast cancer death rates in Hiroshima and Nagasaki. In each separate study the evidence was found to be strongly in favor of supralinearity, meaning that the cancer induction effect *per rem* is higher at low doses than at high total doses. And in complete agreement with these findings, in the BEIR-III Final Report (p. 384) it is demonstrated that exactly the same effect is found in the study of lung cancer in the uranium miners. So all of the evidence is exactly 180° out of phase with the statements about what "the majority of scientists believe," and with what the nuclear promotional community inside and outside of government would like to believe. I am amazed that none of the scientists bothered to check what the evidence indicated before they lent their names to a message of scientific nonsense.

So many people have been brainwashed into believing that *consensus* is what counts that it would not be surprising to see the popularity-contest approach being used to decide all kinds of scientific matters in the future, where vested interests are at stake. The Inquisition proved that consensus rules; witness what they did to Galileo's absurd idea that the sun did not revolve about the earth.

Detoxifying Plutonium as a Pulmonary Carcinogen

The most prominent form of cancer of the lung is that known as bronchogenic cancer, arising from bronchi of intermediate size. Cigarette smoking and exposure to radon decay products in uranium mines are two well-known causes of bronchogenic cancer. The prospect that the inhalation of very fine particles of insoluble compounds of plutonium represents a great hazard for the induction of bronchogenic cancer is one which terrifies those who look forward to a plutonium economy, with plutonium-based breeders and plutonium fuel reprocessing.

Since there have not yet been any adequate epidemiological evaluations of the bronchogenic cancer risk per microgram of plutonium deposited in the vulnerable part of the bronchial tree, except for possibly some crude upper limit evaluations, one is faced with *calculation* as the only way to estimate what the bronchogenic cancer hazard of inhaled plutonium might be. The crude upper limit evaluations can be made from the observations on Manhattan Project workers reported by Dr. Louis H. Hempelmann and Dr. George L. Voelz and co-workers of the Los Alamos Scientific Laboratory.[49] Before considering the calculational estimates, we must help the reader understand *why* the requisite epidemiological studies are not available. For help in this understanding, I quote Dr. George L. Voelz who, to his credit, *has* urged such studies for some time now:

It would be nice to be able to report that the long-term studies on plutonium workers have been practiced faithfully throughout the industry. Unfortunately, the follow-up of workers following termination of their employment in plutonium work has been limited to only a few special situations.[50]

Dr. Voelz has a most charming mastery of the art of understatement.

Until 1975 a marvelous solution was available to the nuclear industry for this distressing problem of induction of bronchogenic cancer by plutonium particulates. There simply could not be *any* bronchogenic cancers of the lung produced by plutonium, *by fiat!* Where did such a remarkable fiat come from? It came from the work of the Task Group on Lung Dynamics of the ICRP, in which it was stated that the retention of plutonium particulates in the bronchi for any appreciable length of time would be *zero*. Quite obviously, since the fiat stated there was going to be no retention of plutonium in the regions of the bronchi vulnerable to the development of bronchogenic cancer, there simply were going to be *no* bronchogenic cancers caused by inhalation of plutonium in the form of insoluble particles. There could not be a better solution for this thorny public relations problem. After all, who could hope for a result lower than zero bronchogenic cancers?

The only difficulty was that there were no data on humans which backed up this claim of zero long-term retention of plutonium in the vulnerable regions of the bronchi. What trivial studies there were on particulate inhalation in humans simply did not *begin* to address the question of long-term retention of particulates in the bronchi.

In 1975 I challenged this *assumption* of zero retention,[51] since the Task Group had based its assumption on the *hope* that ciliary* action and mucus flow in the bronchi would clear out any plutonium particulates. But I cited the extensive evidence that cigarette smokers had severely injured cilia in the bronchial cells, and beyond that, a large fraction of the cilia were actually missing from cells in the bronchi of smokers. I proposed, therefore, that a small fraction of the plutonium retained in the bronchi as a result of ciliary absence and injury could give so high a radiation dose to the bronchial lining cells of cigarette smokers that 0.058 micrograms of plutonium retained by a cigarette smoker for long-term clearance, i.e., half of the plutonium particles expelled in 500 days, would be enough to guarantee a fatal bronchogenic cancer (500 days is the clearance half-time suggested by the Task Group for the nonciliated region of the lung).

The ERDA was very distressed by these proposals of mine and managed to get five of its laboratories to write refutations of my work,[52] all five of which failed to present anything substantive in refutation, as I showed.[53] For example, as supposed evidence that I must be wrong about the long-term retention of plutonium particulates in the lungs of smokers, the short-term studies of particulates by R. E. Albert and co-workers were cited by the ERDA laboratory groups.[54] I pointed out in 1977 that the work of Albert simply had nowhere near the sensitivity needed even to address the question of retention of the amounts of plutonium under discussion, and moreover those studies were all short-term studies, and hence inapplicable entirely.[55]

I would say that *calculations* of the cancer hazard of plutonium are definitely not a satisfactory resolution of this crucial problem. Epidemiological *evidence* will ultimately resolve the question. Calculation is suggested in the absence of data, and if those calculations indicate that a "plutonium economy" may face some very serious problems, it is irresponsible for nuclear promoters not to be willing to face up to those potential problems.

Conclusion

I have illustrated just a few of the ludicrous positions one can find when the benefit-risk doctrine is applied to minimize risks and to exaggerate benefits. One does not have to discuss whether coal pollution is or is not more of an insult to health than is nuclear pollution, or whether or not coal power is cheaper than nuclear power. Benefit-risk analysis is simply and obviously a violation of every human's right not to be trespassed upon by pollutants of any sort, whether or not damage by the pollutant has been proved. If trespass by pollution is acceptable because harm by the pollutant has not (yet) been demonstrated, then invasion of one's home by trespassers would be equally acceptable provided one cannot prove they harmed members of the

---

*Cilia are small hairs on the bronchial surfaces which expel particulates from the lungs by their wave-like action.

household. It is past time that we realize this and stand up for our rights instead of engaging in absurd discussions of whether we prefer cyanide to hemlock.

As for the economic issues involved, the best way to determine the economical and acceptable source of energy is by the free market. All the scholarly debates about which form of energy is cheapest are of no purpose because the number of economic distortions of energy created by government subsidies and privileges are legion. A real free market in energy, of course respecting individual human rights, is the only way to decide economics—and that means a zero role for government.

There will be those who will say that we live in a democracy and one has the right to show his opposition to a particular program by electing different legislators if the current programs enacted for the "social good" are not acceptable to him. This is the philosophy that what the majority decides to do (through its elected legislators) can be imposed upon the minority even if the individual human rights of the minority are totally violated, such as trespass by pollution accepted by the majority. I reject the idea that majorities, minorities, or dictators have any right to violate an individual's inherent rights for affluence or any other goal. Acceptance of any such rights of majorities or dictators is acceptance of the principle that might makes right, since coercion is clearly and unequivocally involved.

The Future: George Orwell Understated the Case

I hope this discussion will lead the reader to contemplate the implications for the future. The government, by confiscation of huge sums of money by taxation, has the power to decide what subjects shall be investigated in science and other fields, who shall be permitted to do the investigations, what results are desirable, and what results should never see the light of day. Nuclear power is a symptom of a very serious problem: total control of what knowledge is generated and is acceptable.

Exaggerating the Risks
BERNARD L. COHEN

A former experimental nuclear physicist turned environmental scientist, Bernard L. Cohen has been a professor of high-energy physics at the University of Pittsburgh since 1958, and was director of its Scaife Nuclear Physics Laboratory from 1965 to 1979. In the 1950s Dr. Cohen worked on cyclotron research at the Oak Ridge National Laboratory. He is the author of numerous scientific papers and several popular books, including *Nuclear Science and Society*. In 1981 he was the recipient of the American Physical Society's prestigious Bonner Prize for nuclear physics, and is currently chairman of the American Nuclear Society's Division of Environmental Sciences. Dr. Cohen, who is an independent scientist with no connection to the nuclear industry or government nuclear establishment, defends the scientific

establishment, arguing that the general public has been manipulated on some of the mos
vital issues of our time by irresponsible journalists and sensation-seeking scientists.

What is radiation, and how dangerous is it? Radiation consists of subatomic particles
that shoot through space at enormous speeds, up to 100,000 miles per second. They
easily penetrate deep inside the human body where they can strike and damage biological cells, and thereby cause cancer or genetic defects in later generations.

This sounds frightening, and one can easily get the impression that to be struck
by one of these particles would be a tragic event. However, this cannot be so, because
each of us, and every other person who has ever lived or ever will live on this earth,
is struck by about 15,000 of these particles of radiation—from natural sources—
every second of his life.[56] In addition, when we get a medical x-ray we are struck by
about a hundred billion of them. What saves us is *not* that it takes some minimum
number of these particles to do us harm; as some like to say, no level of radiation is
perfectly safe. Any single one of these particles can cause a fatal cancer or a genetic
defect in our progeny, but the probability that it will do so is only one chance in 30
quadrillion (30 million billion).[57]

If we are worried about this risk, there are many things we can do to reduce our
exposure. We can wear metal-lined clothing to shield us, like the apron the dentist
often lays across us when he takes an x-ray. We can choose the building materials of
our homes; brick and stone contain more radioactivity than wood and therefore expose
us to more radiation. We can choose to live in areas where natural radiation is lower;
in Colorado radiation is twice the national average whereas in Florida and the Gulf
Coast area it is well below the average.

But most of us do not worry about such things. We recognize that life is a series
of risks. Every breath of air may carry a germ that will cause fatal pneumonia, but
we continue to breathe. Every bit of food may have a chemical that will give us
cancer, but we continue to eat. Every time we get into an automobile we recognize
that we may be killed in an accident, but still we drive. We are willing to engage in
these games of chance as long as the odds are heavily in our favor. And in the case
of radiation we should recognize that 30 quadrillion to one are pretty good odds.
Unfortunately, however, that point is missed by some sensation-seeking writers who
have produced pages of prose about the horrors of being struck by a particle of
radiation.

In evaluating the dangers of radiation, two things must be kept in mind. First, it
is important to be *quantitative,* not just qualitative. With qualitative reasoning, almost
any human activity can be shown to be harmful; there is a widespread impression
that using energy is dangerous and doing without energy improves safety. This is
definitely not so. Only on the basis of quantitative reasoning can rational decisions
be made. The second consideration is to keep things in perspective—we must not
focus on one danger while ignoring all others. With these guidelines, let us proceed.

A radiation exposure of 1 millirem has about one chance in eight million of caus-ing a fatal cancer, and about an equal chance of causing a genetic defect in later generations. These are the estimates by the U.S. National Academy of Sciences BEIR Committee, UNSCEAR, and the ICRP, and they are used throughout the world by all national, state, and local organizations charged with setting radiation standards, including the U.S. Environmental Protection Agency, the U.S. Nuclear Regulatory Commission, and the U.S. National Council on Radiation Protection and Measure-ments (NCRP), etc.

Antinuclear activists charge that industry and government are deceiving the public by concealing the dangers of radiation. But the BEIR Committee is appointed by the National Academy of Sciences, which is composed of about a thousand of the most distinguished scientists in the United States, nearly all of whom are academics with no government connection; the BEIR Committee itself is made up of twenty-one distinguished American scientists recognized in the scientific community as experts in the field of radiation biology, thirteen of whom are university professors; UNSCEAR is made up of scientists from twenty nations from both sides of the "Iron Curtain" and the Third World; and the nine nations with current representation on the Main Commission of ICRP are similarly distributed. It is extremely unusual for such highly reputable scientists to practice deceit, if for no other reason than that it would be easy to prove that they had done so and the consequences to their scientific careers would be devastating. All of them have reputations such that they could easily obtain a variety of excellent academic positions independent of government or industry financing, so they are not vulnerable to economic pressures. But above all, they are human beings who have chosen careers in a field dedicated to protection of the health of their fellow human beings; in fact most of them are M.D.s who have foregone financially lucrative careers in medical practice to become research scientists. To believe that all of these scientists are somehow involved in a sinister plot to deceive the public is indeed a challenge to the imagination.

In quoting a value for the risk per millirem, we are tacitly implying that the risk increases *linearly* with the dose. This is a highly convenient assumption because there is a great deal of information on effects at high doses from a wide variety of sources. These data are all in generally good agreement on the effects of high-level radiation, so if linearity is accepted, effects of low-level radiation are readily calculated by simple proportionality with dose.

There is some question as to whether this simple proportionality with dose can be extended to low levels. The majority of the radiation biology community and all of the prestigious committees believe that this is a "conservative" procedure, more likely to overestimate effects at low levels. The evidence that the linear hypothesis *over*estimates effects at low doses comes from various sources:

1. There is abundant evidence that nature provides mechanisms for repair of radiation damage. First, it has been observed that a given dose of radiation is much less

harmful when spread out in time than when given rapidly,[58] which implies that damage is being repaired in the first case. Second, chromosomes broken by radiation have been microscopically observed to reunite into a single strand.[59] Third, there is well-established evidence for DNA repair in bacteria.[60] The body should thus have little difficulty in repairing the damage from low doses.[61]

2. The great majority of animal experiments indicate that cancer incidence at low doses is substantially less than predicted by linear extrapolation of high-dose data.[62]

3. Data on leukemia among those exposed in the Japanese A-bomb attacks[63] indicate far fewer cases at doses below 100 rem than are expected from a linear extrapolation of data at higher doses, even with generous allowance for statistical fluctuations.

4. Among the U.S. women who ingested radium while painting numerals on watch dials, the number of cases at low doses is substantially less than predicted from linear extrapolation of high-dose data.[64]

All of these lines of evidence indicate that low levels of radiation should be less harmful than predicted by linearity. Nevertheless, until recently most prestigious groups condoned the use of linearity to estimate effects of low levels, and we will use it here.

Radiation Risks in Perspective

Since 1 millirem is a typical radiation exposure in highly publicized incidents—for example, the average exposure received by nearby citizens in the area of the Three Mile Island accident in Harrisburg, Pennsylvania, was 1.2 millirems—let us pause to give some perspective on the dangers of 1 millirem exposure. As already noted, such exposure has one chance in eight million of causing a fatal cancer, which corresponds to reducing life expectancy by 1.1 minutes. This is the amount of life expectancy we would lose from taking three puffs on a cigarette, eating ten extra calories (e.g., one lick on an ice cream cone) if we are overweight, or being exposed to typical city air pollution for one week[65] (see Table 2–5).

In addition, 1 millirem of radiation can cause genetic defects in our progeny. A continuous exposure of the entire U.S. population over several centuries to an additional 1000 millirems up to the age of child conception would increase the frequency of genetic defects by 0.2 percent.[66] Since the average age of conception is thirty years, this is equivalent to exposure to 33 millirems per year. Continuous exposure to 1 millirem per year would therefore increase genetic defects by (0.2 percent/33 = 0.006 percent.

Ordinarily genetic defects arise from spontaneous mutations in the sex cells, which are essentially chemical reactions. It is a well-known fact that increasing the temperature accelerates chemical reaction rates, and that the rate of spontaneous mutation increases very rapidly with temperature. It is widely believed that the anatomy of the

TABLE 2–5   THE RISKS WE TAKE

Activity	Shortened Life Span (in minutes)
Drinking a diet soft drink	0.15
Crossing the street	0.4
Being exposed to 1 millirem of radiation	1.5
Smoking a cigarette	10
Drinking a nondiet soft drink	15
Eating a calorie-rich dessert	50
Flying coast to coast	100
Driving coast to coast	1000
Skipping annual Pap test	6000
Choosing Vietnam army duty	600,000

Source: B. Cohen and I. Lee, "A Catalog of Risks," *Health Physics* 36 (1979): 707–22.

male—with the gonads outside the body—is evolution's way of reducing their temperature and thereby reducing the frequency of genetic defects. The wearing of pants largely frustrates this process, raising the temperature of the gonads by about 3.3° Centigrade, and it is widely believed that this contributes substantially to genetic defects in humans. Judging from animal studies, scientists estimate that this heat doubles the rate of mutations,[67] but even if we assume that it increases this rate by 15 percent, charging this to the typical 5000 hours per year of wearing pants, each hour per year of pants-wearing increases genetic defects by (15 percent/5000 =) 0.003 percent. We thus see that the genetic risk from 1 millirem of radiation is equivalent to about two hours of wearing pants.

The popular notion that genetic effects can do long-range harm to the human race is absolutely false. Genetic selection, otherwise known as evolution, acts to breed out bad mutations and breed in good mutations, so the long-term effect of additional radiation is bound to be *favorable,* rather than unfavorable, because as the centuries go by good new traits are bred in and bad new traits are bred out. Genetic effects of radiation are considered to be bad because in the short term (i.e., over a few generations) they cause birth defects. However, the long-term effect of radiation is negligibly small, and represents no danger to the human race.

Another perspective on genetic effects of radiation may be derived from studies of the Japanese A-bomb survivors. There were about 24,000 in the group who were exposed to an average of 130,000 millirems each, but in the first generation of children born to these people there is no detectable excess of genetic defects.[68]

Irresponsible Reporting

With this background in understanding the risks from radiation, let's review some of the highly publicized incidents involving radiation to the public. By far the most serious was the Three Mile Island accident in which two million people living within fifty miles of the nuclear power plant received an average exposure of 1.2 milli-rems.[69] On an average each of them has an increased probability of dying of cancer equal to one chance in eight million. However, since only two million people have been exposed to this risk, there is only one chance in four of even a single resulting death.

There have been scare stories printed about radioactivity leaks from waste burial grounds in Kentucky and New York. The *Philadelphia Evening Bulletin,* for exam-ple, ran a three-part series with headlines proclaiming "It's Spilling All Over the U.S.," "Nuclear Grave Is Haunting Ky.," and "There's No Hiding Place." But from neither of these leaks were there *any* exposures as large as 0.1 millirems, which is ten times smaller than the Three Mile Island exposures. Moreover, there were only a handful of people involved, so clearly no cancers are expected.* Part of the reason is that only a tiny fraction of the buried radioactive material leaked out. But even if *all* of the radioactivity in those burial grounds were to leak out of the trenches and spread through the soil, there would probably never be a single fatality.[70] The reason for this is that it is highly unlikely that the radioactive waste will ever find its way into our food, even if the land is used to grow food crops. A typical atom in the top layers of soil has only about one chance in a billion each year of getting into human food, which is why it is considered acceptably safe to bury short-lived low-level radioactive materials in shallow trenches. A similar consideration applies to radio-activity released into the air; a particle released in a city has only about one chance in 100,000 of ever being breathed in by a person.[71]

There has been wide publicity of leaks in high-level liquid waste underground storage tanks at the Hanford Atomic Works in Washington State. These tanks were constructed with an obsolete World War II technology and used for military waste. In any event the leaking material has moved only a few feet through the highly absorbent soil, and since the material is forty feet underground with no apparent way to get to the surface or into the river before the radioactivity decays away, it is most difficult to see how anyone can ever be exposed to radiation from those leaks.

Newspapers frequently print stories about transport incidents involving radioactiv-ity in which a package falls off a truck, or leaks a small amount of liquid onto the road. There have been about a hundred such incidents over the last thirty-five years. But these packages carry only small quantities of weakly radioactive material; high-level waste is now carried in special shipping containers which have never broken open or leaked. Typical exposures from these accidents are less than 1 millirem, and there have never been exposures higher than 10 millirems to the public.[72] Since only a very few people are ever exposed in these accidents, calculations indicate that there is less than a 1 percent chance that there will ever be a single fatality from all of these

---

*Even if people received as much as 0.1 millirems, their chances of contracting cancer would be one in eighty million. But clearly eighty million people were not exposed. The fewer people exposed, the less chance (statistically) of a fatal cancer.

accidents combined. Even if a total of 10,000 people has been exposed to an average of 1 millirem, the probability of a single eventual fatality from all of these accidents combined is only $10,000 \div 8$ million, or 0.13 percent. Even this is probably a gross overestimate, since the number of people exposed is probably more like 1000.

Yet in spite of this near-perfect record, the public views radiation as a major threat to its safety, and has been driven "insane" over fear of radiation. How did this peculiar situation come about?

One reason is overcoverage by the media. I did a study of entries on accidents in the *New York Times* Information Bank, a cross-section of all subjects that have appeared in the *New York Times* (and other major newspapers), over the five-year period 1974–1978. There was an average of 120 entries per year on accidents with motor vehicles, which kill 50,000 people annually;[73] on industrial accidents which kill 4500, there were 25 entries per year. But on accidents involving radiation, there were 200 entries per year, more than for the other three types combined, even though there was not a single fatality observed or ever expected from the radiation doses received from any radiation accident during that time period.

In addition to overcoverage the media tend to use inflammatory adjectives like "deadly" radiation, "lethal" radioactivity, whenever they refer to radiation. These adjectives are not applied to accidents involving motor vehicles, or to ordinary air pollution which kills 10,000 people each year, or to electricity which kills well over a thousand by electrocution.[74] With over 100,000 people a year killed in accidents,[75] it is difficult to understand all this attention to a type of accident which is not responsible for a single one of these deaths.

By and large the public believes that nuclear energy is far more dangerous than coal burning. In fact there have been well over a dozen scientific studies comparing nuclear energy and coal, and all conclude that nuclear is safer than coal.[76] (This does not include effects of radon from uranium mill tailings, since the tailings piles are now being covered.) Even Ralph Nader has admitted to me privately that nuclear is safer than coal.

However, these studies are not featured in the media, nor do we hear about the 10,000 fatalities per year—more than one every hour—as a result of toxic chemical releases, better known as air pollution, from coal-burning plants.[77] This means that every time a coal-burning power plant is built, about 1000 people are condemned to an early death from air pollution. By comparison even the antinuclear activists do not claim that there will be more than 100 excess fatalities from a nuclear power plant. Failure to inform the public on this point is surely a grave irresponsibility which, over the years, will cost our country tens of thousands of lives and hundreds of billions of dollars.

The media have portrayed the reactor meltdown as the "ultimate disaster," and the Three Mile Island accident is frequently referred to as a "close call." On the rare occasions when media people discuss the actual consequences of a meltdown, they refer only to the worst meltdown, a type expected once in 100,000 meltdowns. (And we must not forget that we have not had one meltdown.) The NRC estimates that most meltdowns would cause no fatalities,[78] because the radioactivity would be prevented from escaping by the containment structure, the powerful "fortress" within which all U.S. power reactors are sealed. All official reports on the Three Mile Island accident agree that even if a meltdown had occurred, the resulting health effects would have been relatively inconsequential because there was little threat to the containment vessel. An average meltdown is estimated to cause 400 fatalities,[79] which means that for nuclear energy to match the 10,000 fatalities per year from coal burning, there would have to be a meltdown every two weeks. Even if we use the estimates developed by such nuclear critics as the Union of Concerned Scientists,[80] there would have to be a meltdown every six months to make nuclear energy as dangerous as coal burning.

The media have also led people to believe that there has been much excess childhood leukemia in southern Utah downwind from the Nevada bomb test site, caused by radioactivity releases in those tests. The only scientific evidence on the subject is a paper by Dr. Joseph Lyon (a University of Utah epidemiologist) and collaborators, which offers evidence that there were between two and twenty-five excess cases of childhood leukemia in the region.[81] Their evidence is that the incidence of this disease was higher during the bomb-testing period than before or after (although at all times it was less than the U.S. national average). There are a number of reasons for uncertainty in the Utah researchers' conclusions: (1) there were no excess childhood leukemias among the Japanese A-bomb survivors who were more than about one mile away from the bombs dropped on their cities; (2) there were no excess thyroid cancers in the Utah area, although many more of these cases could be expected than of leukemia from bomb fallout; (3) the same statistical analysis used to demonstrate excess childhood leukemias shows a *deficiency* of all other types of cancer during the sensitive time period, a result which is as statistically significant as the excess of leukemias; (4) "clustering" of leukemia cases at certain places and times has been observed in other situations with no connection to radiation (this is sometimes cited as evidence that a virus is involved); (5) Lyon himself admits he is far from certain about the effect, and at least two papers by other scientists have expressed deep skepticism; and (6) the radiation doses are highly uncertain, thus providing little quantitative information about health effects of radiation. Nevertheless television specials have focused on excess cancer in the Utah area, and pointed to various types of adult cancer, for which there is no evidence, as established fact. Unfortunately, and erroneously, the impression has arisen that there have been large numbers of fatalities.

A few people in the scientific community have challenged the generally accepted position that if all our electricity were generated from nuclear energy, the increased annual radiation dose to the public would be about 0.2 millirems per year. The most highly publicized study, published in 1971,[82] was one which assumed that the average American would be exposed to 170 millirems per year because that is the internationally accepted *maximum* permissible dose to the general public. The study gave no explanation for how a person on the average can be exposed to 170 millirems per year when federal regulations require that no single member of the public can be exposed to more than 5 millirems from nuclear plants (the latter requirement assures that the average citizen is exposed to no more than about 0.2 millirems per year, as was stated earlier). Thus the study's doses were nearly 1000 times (170 ÷ 0.2) too large. In addition the study used risk estimates five times larger than those accepted by all the prestigious committees and standard-setting bodies, so the study's estimates of fatalities from the nuclear industry are about 5000 times too high.

The other challenge was from a recent report published by an antinuclear group in Heidelberg, which claimed that German and United States government dose estimates from routine radioactivity releases from nuclear plants underestimate the effects of radiation by nearly 100 times. The report[83] was based entirely on theoretical calculations; the group never conducted actual measurements of radiation in the vicinity of any nuclear power plants. They predicted very high radioactivity levels on the ground, in the air, in vegetables, and in milk. However, these levels are measured routinely by federal and state officials, as well as by the plant operators and by independent research scientists like myself, and they are found to be very much lower than the Heidelberg estimates.

Neither of these studies had much impact on the thinking of the scientific community, which views them with extreme skepticism, but needless to say they are still accepted and used by some segments of the antinuclear movement.

Plutonium Toxicity

One of the worst distortions of radiation dangers involves the toxicity of plutonium. Plutonium is basically no different from many other radioactive materials, but somehow it has been singled out for special treatment by some antinuclear activists who put forward the so-called hot particle theory.

Indeed the "hot particle" theory had been looked into by reputable scientists from time to time over the years, so when a group of antinuclear advocates raised the issue in a 1974 legal petition,[84] several investigating bodies consisting of scientists specializing in that field were set up: a committee of the National Academy of Sciences, groups from the NCRP, the British Medical Research Council (MRC), the United Kingdom National Radiological Protection Board (NRPB), and a study by three eminent scientists instigated by the U.S. Atomic Energy Commission. All of these studies concluded that there is no merit in the "hot particle" theory; that concentration of radioactivity in particles is, if anything, less dangerous than spreading it out uniformly over the lung.[85] As a result of these reports the scientific community has ignored the "hot particle" theory, and the standard-setting organizations did not revise their standards as they would have been forced to do if any credibility were attributed to that theory. Nevertheless the antinuclear activists continue to use the theory to justify such quotes as "a single particle of plutonium inhaled into the lung will cause cancer," "plutonium is the most toxic substance known to man," and "a microgram of plutonium will cause cancer."

In 1976 I published a paper on plutonium toxicity in which I used the procedures recommended by all of the standard-setting committees to derive practical estimates of health effects in various scenarios involving plutonium dispersal.[86] The principal conclusions of my paper were that, on an average, about 200 micrograms of plutonium inhalation would be needed to cause a lung cancer, and that if a pound of plutonium were released in a large city in the most effective practical way, the total number of expected fatalities would be about twenty-five.[87] These conclusions took into account plutonium's long-lasting effects over thousands of years. An important point here is that when material is dispersed into city air, only about one part in 100,000 finds its way into human lungs.

Ralph Nader asked the NRC to review my paper. Judging by the number of telephone calls I received from NRC personnel asking about calculational details, many man-days of effort were devoted to the review, but in the end it was given a clean bill of health. Nevertheless, Nader accused me of "trying to detoxify plutonium with a pen."[88] My response was to offer personally to inhale 1000 particles of plutonium of any size that could be suspended in air, or 10 micrograms of plutonium in any form (ten times the dose that Nader claimed would be lethal). I sent my offer to all major television networks, the "60 Minutes" program, and Johnny Carson, asking that I be given a few minutes to explain why I was doing this: my risk, according to my calculation, was equivalent to the risk faced by a soldier in wartime, and I felt that my demonstration would be an important service to the country. My offers were ignored, but I hope Nader now recognizes that I am not merely "trying to detoxify plutonium with a pen."

After the fuss about the "hot particle" problem had died out, another study appeared, proposing a new and different theory of why plutonium is much more toxic than given by standard estimates like mine.[89] The study assumed that smokers are more susceptible to plutonium than nonsmokers because their lung mechanism for clearing out dust particles that settle on the bronchi is damaged.[90] However, direct experimental measurements of the clearance times for dust to flush itself out of the lungs show that there is essentially no difference between smokers and nonsmokers. Smokers compensate for loss of normal clearance mechanisms by extra coughing and mucus flow; if they actually cleared their bronchi as slowly as the study assumed, they would die of suffocation. In addition the study used a crude estimate of the surface area of the bronchi, which is seventeen times smaller than the generally accepted area. The study also used the BEIR Report's maximum estimate of risks rather than its best estimate. The study concluded that plutonium is a thousand times more dangerous for smokers than usual estimates, including mine.

Simultaneously another paper appeared, also not published in any scientific journal, which estimated that in a breeder-reactor economy there would be 20,000 fatalities per year from plutonium.[91] The paper began with a statement that sounds reasonable: that the nuclear industry cannot contain more than 99.99 percent of its plutonium. However, the paper then *assumed* that all of the remaining 0.01 percent will be dispersed into the atmosphere as dust so fine that the particles will remain suspended in the air for at least several days—it is only these very tiny particles that can get past the body's defenses into the deep lung where they can do damage. However, only a minute fraction of lost material is likely to escape as dust, so it is misleading to assume that *all* of it will.

Achieving 99.99 percent containment sounds difficult. In reality virtually all industries prevent many times less than 0.01 percent of their material from escaping as fine dust. Even the steel and asphalt industries achieve 99.999 percent containment,[92] in spite of the fact that they heat their product far above the melting point, resulting in vigorous bubbling and boiling, which is an ideal way to create tiny particles suspended in air. On the other hand plutonium for breeder reactors is never melted, let alone boiled. Steel and asphalt are handled in the open, allowing easy escape of these particles, whereas plutonium is always tightly enclosed in a box within a box, so the outer layer maintains containment even if the inner one should break open. Moreover the entire system is always inside a sealed building (or truck) from which air can only exit through filters capable of removing plutonium dust. Present federal regulations require that releases be kept to about one part per billion, or 0.0000001 percent containment, and present plants are operating within those regulations which correspond to 100,000 times better containment than 0.01 percent.

There have been serious accidents, including a fire at a Rocky Flats, Colorado, plant in 1957 in which 0.0003 percent of the plutonium that burned was released, but consequently many improvements were designed into plants to reduce this problem. As a result in a 1969 fire which did $50 million in damage, only 0.000003 percent of the involved plutonium escaped. Of course accidents are rare events, and the vast majority of plutonium is processed without ever being involved in an accident. Furthermore, breeder-reactor plutonium, unlike the plutonium processed for weapons at Rocky Flats, cannot burn.

The health effects of radiation are strictly scientific questions about which there is very little controversy in the scientific community. Nevertheless a small group of dissident scientists—fewer than ten with scientific credentials in the field—have tried to convince the public that dangers of radiation have been underestimated. They have received tremendous support from the media, which have neglected to report on the rest of the scientific community. Since scientists have little opportunity to communicate with the public other than through the media, it is no wonder that the public has become confused. This essay is an attempt to straighten out some of the confusion.

3

Reactor Safety

Introduction

I do not think that it is presently worthwhile making a laboratory out of the United States so these people can learn how to deal with reactors.

—Dr. Henry Kendall, professor of high-energy physics at MIT and chairman of the Union of Concerned Scientists

Industrial reactors are so safe that we know of not a single individual in America whose health has been damaged by the nuclear nature of such a reactor.

—Dr. Edward Teller, father of the hydrogen bomb and first chairman of the Atomic Energy Commission's advisory committee on nuclear safeguards

Three Mile Island demonstrated to the world the potential for destruction that is locked inside every nuclear power plant. But we live daily with hazardous technologies where the likelihood of a disaster is perhaps greater than with nuclear plants—risks from coal and chemical plants, to name a few. And most people would claim that these are technologies we cannot do without. Are the risks from nuclear power comparable to those from other technologies which modern civilization has learned to tame, or is nuclear technology, with its awesome potential for disaster, qualitatively different? Moreover, if nuclear power is not 100 percent safe, is it safe enough? These are a few of the questions that enter into the reactor safety debate.

Ten billion curies of radioactive fission products are locked inside a typical reactor core—stored inside a 400-ton steel reactor vessel which in turn is housed inside a containment building made with reinforced concrete. Cooling water circulates through the core to carry away the energy generated by the chain reaction. The cooling water also prevents the core from heating up past the melting point of uranium dioxide (5000°F) and keeps the radioactive fission products contained in the reactor vessel.

Perhaps the most serious danger for a nuclear reactor would be a loss-of-coolant accident (LOCA). The flow of coolant water to the core could be interrupted by any number of failures: if one of the main pipes carrying water into the reactor vessel cracked under pressure because of a faulty weld, or if a stuck valve went unnoticed by plant operators (which is what happened at Three Mile Island). If a LOCA went unchecked, a meltdown of the uranium core could lead to disaster.

To prevent this engineer's nightmare, every reactor is equipped with layers of elaborate, redundant safety systems. The emergency core cooling system (ECCS), for example, is designed to dump hundreds of thousands of gallons of cooling water automatically onto the exposed core in a matter of minutes. But what would happen in the unlikely event that, during a LOCA, the ECCS for some reason failed to kick on or was turned off manually?

A meltdown could possibly occur. A probable scenario is as follows: within minutes most of the cooling water would rapidly pour out of the vessel and spill onto the reactor floor, exposing the twelve-foot uranium core. Without water, the chain reaction in the core would come to a complete halt.* However, the decay heat of the fission products would continue to generate large amounts of heat for days. Within the next few hours, if operators failed to turn on the ECCS, the 100 tons of uranium could melt through the vessel's eight-inch-thick carbon-steel walls and slump onto the concrete basement floor of the containment building. Pressures could build up to dangerous levels from the hot gases forming in the building: steam would be created from the molten uranium dropping into the coolant water, carbon dioxide gas would be released from the disintegrating concrete basement, and hydrogen gas would be created from the zirconium in the fuel rods interacting with water. If the pressures exceeded 100 pounds per square inch (seven times atmospheric pressure), the dome of the reactor might possibly crack. Or if the molten uranium plunged into a pool of water, a rapid pressure rise or steam explosion might possibly rupture the walls of the containment building, allowing hot radioactive gases to escape into the atmosphere.

The results could be deadly. A study conducted by the Atomic Energy Commission (AEC) in 1957 acknowledged that a major accident could affect an area roughly the size of the state of Pennsylvania. However, it is expected that only a small number of meltdowns—barely 2 percent—would actually breach the containment building and lead to a catastrophic release of radiation. Furthermore, because the chances of *any* meltdown's occurring are seen to be small, most scientists conclude that the odds are greater that one will die from coal pollution, from a dam burst, or from a gas explosion, than from a meltdown. Nuclear engineers point to several reassuring facts:

○   According to sophisticated computer studies, the probability that the ECCS would fail to deliver water on demand is less than one in a thousand LOCAs.

○   In 1980 scientists at the Sandia Laboratory in New Mexico deliberately attempted to create a large steam explosion by dropping molten uranium into water and were unsuccessful.

○   In December 1978 scientists at the government's National Reactor Testing Station

---

*The neutrons generated by the fissioning process are unable to sustain the chain reaction without water. Collisions with water molecules slow down the neutrons and increase the rate of fissioning.

in Idaho Falls simulated the worst possible LOCA with the Loss of Fluid Test (LOFT) reactor and found that the ECCS delivered water *sooner* than the calculations had predicted. (Although the demonstration reactor was only one-fiftieth the size of a commercial 1000-megawatt PWR, it was operated at the same temperatures and under similar pressures to those in reactors operating today.)

Many critics agree that a nuclear power plant operating in top form would probably be relatively safe. But they are convinced that there is a wide gap between theory and practice; some reactors are poorly designed or constructed, others are badly managed, and the prospect of a tired or incompetent operator at the controls is frightening. According to the NRC there were 3804 small safety-related incidents involving reactors in 1980 alone. Most were no more dangerous than a gummed-up control rod or a hairline turbine crack. Nevertheless several minor incidents could trigger an unpredictable cascade of failures like the accident at Three Mile Island which, some assert, came uncomfortably close to releasing extremely large amounts of radioactivity.

Throughout most of the 1950s and 1960s the safety of nuclear reactors was rarely questioned. Few reactors had yet gone on line, and experimental safety tests at the LOFT reactor in Idaho Falls were due to begin in 1966. (Because of federal budget cuts and conflicts of interest, the tests did not begin until 1978.) By the late 1960s the AEC had invested tens of millions of dollars in reactor safety research. Nuclear critics were viewed as noisy radicals who were protesting the establishment more than the safety of reactors.

By 1970, however, questions were being raised about the reliability of the computer calculations used to predict the behavior of the ECCS; in a series of small mock-up tests in 1970–1971 in Idaho Falls, the ECCS sometimes failed. In February 1972 long and acrimonious congressional hearings began investigating the industry's claims concerning the effectiveness of the ECCS. During the hearings nuclear safety experts from within the AEC testified that technical problems existed, and that the AEC had suppressed unfavorable safety reports. As a result, supplementary safeguards were introduced—enough to satisfy the engineers and scientists who had testified.

The reactor safety controversy flared again in 1974 when the AEC released the results of a $3-million safety study, the Rasmussen report (often referred to by its AEC number, WASH–1400). Although Dr. Rasmussen calculated that the very worst accident would be devastating—3300 people would be killed immediately, 45,000 would probably die later of cancer, and an area of several thousand square miles would be contaminated for years—he also concluded that the likelihood of such an uncontrollable meltdown was *once in a billion reactor-years.* *

* A reactor-year is one reactor operating for one year; ten reactors operating for five years would constitute fifty reactor years of operation. The report also concluded that a much smaller meltdown could occur once in 20,000 reactor-years. Before the Rasmussen report, physicists assumed that a serious reactor accident could occur once in a *million* years of reactor operation.

The Rasmussen report was criticized on several accounts: not only did the summary underestimate the number of latent cancer deaths, but the way in which the probabilities were calculated allowed for too many uncertainties. In light of the criticism the NRC asked a panel of seven experts to review the report. The chairman was Dr. Harold W. Lewis, a pro-nuclear University of California physicist who is a stickler for safety. While the Lewis report criticized Rasmussen's summary, it acknowledged the study as a milestone in reactor safety research. One member of the panel, Dr. Frank von Hippel, a physicist at Princeton University who calls himself "anti-pro-nuclear," summarized the panel's conclusions when he said, "All of us think reactors are safe enough so that you can't use safety as a reason to reject nuclear power."

Whether or not a computer calculation would indeed provide a reliable estimate of the ECCS's performance did not seem to be of concern to the general public, perhaps because lay people still felt ill-equipped to judge the intricacies of the technology. Three nuclear engineers did attract some attention in 1976 when they resigned from General Electric in protest of unresolved safety issues, but for the most part the debate over reactor safety caused only mild reverberations.

All this changed on March 28, 1979, when a faulty valve along with human error in Harrisburg, Pennsylvania, triggered a ten-hour accident that may take ten years to clean up. Overnight nuclear terms like "meltdown," "reactor core," and "control rods" became part of the American vocabulary. Hundreds of reporters swarmed to the scene at Three Mile Island to determine what went wrong.

Critics maintain that the accident was a close call, and that next time we will not be so fortunate. Indeed, according to the Rogovin report, which was sponsored by the NRC to investigate the accident, the reactor core came within 30–60 minutes of melting down. The advocates reply that the accident demonstrated the strength of the safety systems: nearly every possible abuse was hurled at the reactor core and still it did not melt down.

Both sides agreed that communication could have been better between the industry and the NRC. An incident similar to that at TMI had occurred at the Davis Besse reactor in Ohio in 1977. An accident was averted, but the engineer's report, stressing the need to take serious precautions to avoid a similar mishap, was never circulated.

As of January 1982, seventy-two licensed nuclear reactors were operating in the United States, seventy-eight had been granted construction permits, and construction permit approval was pending for eleven more. In more than twenty-five years of commercial nuclear power, the industry had compiled an impressive reactor safety record: only one major loss-of-coolant accident, and no lives lost. Safeguards at nuclear plants are more stringent than those for any other energy technology. Yet the question remains: Is nuclear power a safe technology?

The advocates believe that it is. "Nuclear power," said one engineer, "is a low-risk, high-dread technology." To the defenders of nuclear power the potentially catastrophic effects of a major meltdown have no meaning unless probabilities are attached to the calculation. In their opinion the critics have unduly alarmed the public by exploiting the worst-case scenario while ignoring the *likelihood* of such an event: once in a *billion* reactor-years. Pro-nuclear engineers find probability calculations to be sufficient indication that the risks imposed by nuclear reactors are no worse, and probably fewer, than those we accept routinely from airplane crashes and a host of natural disasters. "People resent probabilities," said Dr. Lewis, "but it's only fair to deal with these things quantitatively. Those who say 'there shouldn't be one life lost' may enjoy the chest beating, but it's just plain silly."

The critics would agree that any technology involves risk. However, they are convinced that the effects of a major meltdown are far more devastating—physically, psychologically, and economically—than the effects of coal pollution, airplane crashes, or dam bursts. In their opinion the problem lies historically with the fact that the nuclear establishment, eager to develop a thriving industry, failed to acknowledge the qualitative difference between nuclear power and other energy technologies. Not only did the industry build nuclear plants near metropolitan areas (and allow the Indian Point #1 reactor twenty-five miles north of New York City to run for twelve years without an ECCS), but the government limited the industry's financial liability to $560 million (through the Price-Anderson Act in 1957) despite the fact that an AEC study released the same year estimated that the damage could be as high as $7 billion.

To an extent the nuclear industry has borne the brunt of the public's rude awakening to our modern age. In the 1950s and 1960s people marveled that man could put an astronaut on the moon, and develop transistor and computer technologies. Today the public feels ill-equipped to cope with some of the side effects of modern technology: pesticides in our food, PCBs and other chemicals in our rivers, asbestos in our workplaces, and Love Canal in our memory. In the last decade the reactor safety debate has focused on much more than the reliability of the ECCS or of probability statistics; the controversy has become a sounding board for people's views of modern technology and industrial progress in general.

To the critics nuclear power signifies the abuse of a technology: it is the proverbial genie that cannot be put back into the bottle. Disagreeing with the opinion that nuclear power is simply another engineering challenge to modern man, critics of the technology believe that with nuclear power we have gone too far. Advocates, on the other hand, would agree that nuclear power has been singled out to represent the abuse of a technology, not because the *technology* is dangerous, but because the *critics* are dangerous. Like the Luddites, an irrational group of protestors in England in the early 1800s who broke into textile mills at night and destroyed the machinery, advocates believe that these nuclear Cassandras would have us roll back the wheels of progress. "If you're worried about the dangers," said Dr. Bernard L. Cohen, "then worry instead about people being poor and miserable from lack of energy."

Have scientists devised enough safety features to prevent a catastrophic accident from occurring? Or is there a chance that we will see a meltdown in our lifetime? These and other concerns have made the American public more wary, perhaps, of nuclear power than of any other energy technology. In this chapter two nuclear physicists consider the risks of nuclear reactors. Weighing all safety considerations, they debate the question of whether modern technological society can cope successfully with energy from the atom.

Safe Enough

ANTHONY V. NERO, JR.

Anthony V. Nero, Jr., is a nuclear physicist with the Energy and Environmental Division of the University of California's Lawrence Berkeley Laboratory and the author of a popular introductory textbook on nuclear reactors *A Guidebook to Nuclear Reactors*. He was a research fellow in nuclear physics at the California Institute of Technology from 1970 to 1972 and an assistant professor of physics at Princeton University from 1972 to 1975. While Dr. Nero would agree that the accident at Three Mile Island highlighted many deficiencies in nuclear safety, he believes that improvements made since TMI, combined with sufficient margins of safety in overall reactor design, make nuclear power one of our safer energy options.

Are reactors safe? Or are they only safe enough? Such a fine distinction illustrates an important notion for examining the safety of nuclear power plants. Some opponents of nuclear power claim that reactors are unacceptable because they cannot be made *entirely* safe, i.e., with zero chance of a major accident. In fact until recently some strong advocates of nuclear power claimed that there was *no* chance of a major accident, but events of the last decade have shown this to be untrue. The only practical requirement for nuclear power or any other energy technology is to diminish the chances of a big accident. If careful design and operation make the accident probability tiny enough, then nuclear reactors will be adequately safe.

Even twenty-five years ago a report published by the Atomic Energy Commission acknowledged that an accident in a nuclear reactor could in principle cause substantial damage.[1] But in 1957 the AEC also thought that a core meltdown and major release of radioactivity were so improbable—once in a *million* reactor-years—that the average risk from accidents was virtually zero.

More detailed assessments of reactor safety began in the early 1970s. To the surprise of many in the nuclear business, these studies indicated that accident probabilities were larger than engineers had previously thought. But the risk was still small compared with risks from other energy technologies, and smaller than many risks that the public accepts routinely. Reactors are carefully designed with numerous basic features and safety systems that serve to contain radioactivity, *especially* in an accident. With vigorous safety regulation, accidents will be highly improbable and reactors will be safe enough, even safer than most of the alternatives.

Layers of Safety

The basic reactor system has a number of features that tend to confine radioactivity effectively and to keep the chain reaction under control.[2] These include the *fuel rods,* containing the uranium oxide fuel pellets; the *primary system,* consisting of the reactor vessel and associated pipes and pumps; and the water *coolant* itself. The fuel rods consist of materials with great structural strength: the chemical and physical form of the fuel pellets and the "cladding" that contain them are chosen because of their resistance to corrosion or melting at high operating temperature; the cladding (basically long, thin tubes made of zirconium alloy) is sealed so that even gaseous forms of radioactivity cannot ordinarily escape.

Should some radioactivity escape from the fuel rods, either because of routine leaks in the cladding of a few of the rods or even because of overheating of a large part of the core, the primary system retains the radioactivity until it can be removed by systems that clean the cooling water. The reactor vessel itself is a massive cylinder, weighing hundreds of tons, made of steel eight inches thick, and tested to withstand very high pressures and temperatures, as are the huge pipes and pumps connected to the vessel.

Finally, the coolant itself provides a measure of protection against accidents, beyond its basic function of carrying away the heat generated in the core. Should the reactor lose its cooling water, the chain reaction immediately stops because neutrons are no longer slowed down sufficiently to cause the fissions that continue the reaction.

In addition to these intrinsic safety characteristics, numerous "engineered" safety systems exist to prevent single, or even multiple, failures from turning into serious accidents. In the event of virtually any abnormality, however small, the shutdown control rods can be shoved into the core between the fuel rods to absorb neutrons and turn off or "scram" the chain reaction on a moment's notice. The fact that the scram system is set to respond even to minor changes in operating conditions, such as a sudden shutdown of the turbine generator, or minor changes in the pressure of the cooling water, helps guarantee its function in case of a serious failure.

However, turning off the chain reaction does not reduce energy generation to zero. The large amount of radioactivity in the fuel rods continues to generate heat, albeit at only a few percent of the rate when fission was taking place. But this residual "decay" heat is still enough to melt the fuel rods if cooling is not provided. So reactors are designed for continued circulation of cooling water even after the chain reaction is shut down.

The most serious and fundamental danger for a nuclear reactor is a loss-of-coolant accident (LOCA) whereby, because of some technical failure, the flow of coolant water through the reactor core is interrupted. Because a LOCA can be very serious, every U.S. reactor has an array of emergency core cooling systems (ECCS) that are available to supply cooling water to the primary system. For example, a pressurized-water reactor (PWR) such as the one at TMI has a "high-flow, low-pressure" emergency system—simply, a pressurized water tank that automatically dumps its contents, without pumps, into the reactor if the primary system's pressure drops drastically. A PWR also has emergency pumps that can supply water at higher pressure or lower flow rate. Boiling-water reactors (BWRs) have similar redundant emergency systems.

In the unlikely event that a LOCA occurs and the ECCS fails to operate as designed (or is turned off, as at TMI), the fuel can be damaged severely, even melted. Radioactivity released from the fuel rods is ordinarily retained by the primary system, but if water is leaking from the system the radioactivity could escape from the reactor. However, even if the reactor systems fail, permitting radioactivity to escape the system, a nuclear power plant is equipped with additional features to protect the public against radioactivity: the steel-lined containment building surrounding the reactor provides a substantial line of defense against a massive release of radioactivity. Made of reinforced concrete with an inner steel dome-like shell, this structure is designed to withstand pressure increases from large amounts of steam escaping from the reactor during a LOCA. To keep the pressure down, the inside of the dome is equipped with spray nozzles and/or refrigeration systems that will condense the steam and reduce the pressure.

In addition the containment building has built-in filter systems to remove radioactivity from the air inside the dome. Also, many of the radioactive gases, especially iodine, being cooled by the spray will condense on the many surfaces inside the containment building. At Three Mile Island these mechanisms successfully prevented almost all of the harmful iodine from being released over Harrisburg.

Even the "China Syndrome," popularized by the movie of the same name, would not release radioactivity directly into the air. This somewhat fatuous term describes an accident where the cooling water leaves the primary system and the ECCS fails. The fuel then falls out of the core to the bottom of the reactor vessel, heats up, and melts its way through the thick steel vessel and eventually through the foundation of the containment building. As bad as it sounds, the China Syndrome would have minor health consequences because most of the radioactivity would make its way into the ground or bedrock beneath the power plant, with only minute quantities filtering into the air. Moreover, the fuel would cool down and resolidify within fifty feet below the containment building. There would be no trip to China.

Weighing the Risks

The *probability* of accidents depends on the design of the reactor and associated safety systems. The *consequences* depend on a number of other factors: the amount of radioactivity that is released, the proximity of people to the plant (which would be affected by the amount of warning given), and the weather (which determines how airborne radioactivity is carried from the plant). Reactors are usually built in relatively remote areas as a precautionary measure.

Although the AEC's 1957 analysis was fairly primitive, it recognized that both the probability of accidents and their consequences are equally important in determining the actual risk from accidents. For example, if an accident of a particular type could be expected to kill one thousand people, but the probability of occurrence was only once per million years of reactor operation, then the average risk is one one-thousandth of a death per year of reactor operation. That is, the average risk from a particular accident is obtained by multiplying the consequences of the accident, in this case a thousand deaths, by the probability of recurrence. The average risk is so small, one death every ten years, because the probability of this accident's occurring is only once in a million years.

Thus although the trauma of such an accident would be great, the net risk to society is small compared with other risks that we accept routinely. For example, airline accidents can also kill hundreds of people, and these occur with notable frequency. The consequences of dam failures could be even larger, thousands or tens of thousands of deaths, and these have higher probability than large reactor accidents. Ironically even these large-consequence accidents contribute, on the average, only a few hundred deaths per year of the over-100,000 total accidents suffered by the U.S. public annually. These arise almost entirely from relatively high-probability small accidents: automobile collisions, fires, falls, etc. Even if the average risk from nuclear accidents were as much as ten deaths per year for each reactor, similar to the liability from an ordinary coal-fired plant, this would contribute very little risk to the public.

These basic notions of risk have long been understood by those who examine the liability of energy technologies. But a nuclear power plant is so complex that it has been difficult to understand the risk from reactor accidents quantitatively. The first comprehensive attempt was made in the early 1970s when the AEC commissioned Dr. Norman Rasmussen to identify the various types of reactor accidents systematically, to characterize them in terms of radioactive releases, and to estimate their consequences and probabilities. The quantitative results of the Rasmussen report (also known as WASH–1400)[3] have substantial uncertainties and the report is seriously misleading in some respects. However, the analysis was a watershed in the history of reactor safety and the first substantial step in our understanding the risk from reactor accidents.

The Rasmussen report used probabilistic analysis techniques developed for decision-making in business and for reliability analysis in the aerospace industry to calculate the likelihood of rocket failures. The approach is based on an "event tree" that outlines the possible sequences of events emanating from any specific failure, such as a LOCA due to a broken pipe or a blown seal. Even for a particular failure, these sequences are distinguished from one another by whether or not each of the safety-related systems (e.g., supply of electric power, the ECCS, the containment vessel) functions properly. Once the failure probabilities of each of these systems is known, they can be multiplied together to determine the probability of any particular accident sequence's occurring. If a safety system is particularly complex, its failure probability is *first* calculated by a "fault tree" that uses estimates of the failure rate of specific components of the system and of those who operate it. The components of these systems include ordinary industrial items, such as valves, pumps, and electrical relays.

We turn first to Rasmussen's results, paying particular attention to the deaths that might be caused by the radioactivity released. Fig. 3–1, adapted from the Rasmussen report, shows the probability per year of reactor operation of an accident that would cause fatalities exceeding the number indicated.

Fatalities were broken down into two categories: latent cancers, which would occur years after exposure; and "early" fatalities, which would occur immediately to people exposed to large doses of radiation. In either case the chances of large numbers of deaths are very small: the chance that an accident would cause 3000 latent fatalities is one in *half a million* reactor-years. The chance of an accident's causing the same number of early fatalities is a thousand times less, or one in a *billion* reactor-years. Adding all types of accidents, large and small, for each year of reactor operation, an accident is estimated to cause one-fiftieth of a death from cancer and less than one-ten-thousandth of an immediate death from large radiation exposure.

Their estimates are very uncertain, which the authors of WASH–1400 would be the first to acknowledge, as indicated by the air bars in Fig. 3–1. For example, although their estimates indicate a one in 30 million chance per reactor-year of an accident causing at least 300 fatalities, their figures also calculate that this chance may be high or low by as much as a factor of five, and the number of deaths may be anywhere from 75 to 1200. It is extremely difficult to predict and estimate the probability of human, mechanical, and design failures. One criticism of WASH–1400 is in fact that even these large uncertainties appear to be underestimates.

Unfortunately, the WASH–1400 report presented the work poorly, and the summary is misleading. For example, in discussing the chances of fatalities the summary virtually neglected latent cancers, although these were estimated to be hundreds of times more numerous than early fatalities. Moreover, the body of the report was presented in such a way that other fundamental issues, such as the reliability of the regulatory system for assuring the quality of reactor components, were obscured.

However, even with the uncertainties most analysts—including many who oppose continued use of nuclear power—agree that the average risk from nuclear power plants is no more, and probably much less, than the risk from burning coal. This consensus of sorts arises largely from the major step forward represented by WASH–1400. With all its defects, WASH–1400 still serves as an indispensable guide to the analysis of nuclear accidents, as a basis for regulatory efforts (including those aimed at possible design improvements), and as a foundation for future reactor safety studies.

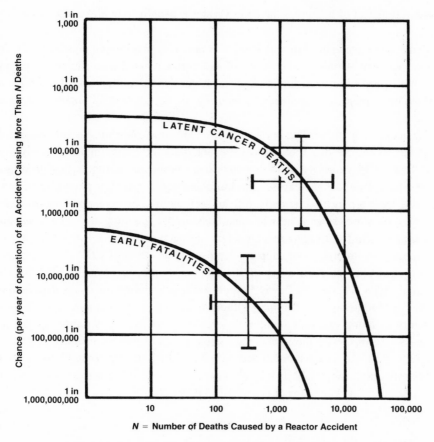

Fig. 3-1 Cancer Deaths from a Major Nuclear Accident. Error bars denote large uncertainties.

Learning from Our Mistakes

While the preliminary draft of WASH–1400 was still the subject of hot debate, an accident occurred that underscored some of the major issues: a fire at the Tennessee Valley Authority's Browns Ferry nuclear generating station in Alabama. In March 1975 this facility had two nuclear plants operating and one under construction. Incredibly, workers were using a candle to test for air leaks in a room for electrical cables, and they accidentally started a fire that destroyed many of the cables controlling safety systems at both reactors. Although both reactors were scrammed within the hour, it took several hours to bring the fire under control. In the meantime water was being supplied to the reactor of Unit 1 by a pump that was not large enough to make up for the water being boiled away by the hot fuel, and the water level dropped to within four feet of the top of the core. (Water is usually kept eleven feet above the core.)

Many critics argue that Browns Ferry was only one small pump failure away from a major accident. However, as the NRC and others pointed out, alternative pumps could have been used to maintain adequate water supply and prevent the core from overheating or melting, had they been required. In any case, at no time was the core exposed.

The accident at Browns Ferry raised the fundamental question: What happened to the quality assurance that should have prevented simultaneous failure of many systems? Browns Ferry uncovered some significant problems, particularly with regard to fire safety. The NRC and those who build and operate nuclear plants now ensure that cable trays are located far enough from each other to prevent simultaneous damage, and look more critically at other quality assurance requirements. As a result, the regulatory system for assuring reactor safety has been strengthened. Even so, lingering doubts about the regulatory apparatus arose in the spring of 1979.

Early that year the NRC accepted the findings of a review panel on risk assessment, led by Dr. Harold W. Lewis,[4] going so far as to withdraw its endorsement of the WASH–1400 executive summary. Interestingly, the panel recommended that the regulatory system give closer attention to small LOCAs caused by minor pipe breaks and leaks. According to the panel, the system was too much concerned with very large LOCAs of the sort that could occur if one of the huge pipes supplying water to a PWR split open completely. It is ironic that the accident at Three Mile Island within a few seconds of its beginning was a small LOCA.

Three Mile Island

In the predawn hours of March 28, 1979, an unremarkable incident occurred at the Three Mile Island nuclear station near Harrisburg, Pennsylvania. This incident and those that followed gave direct and disconcerting evidence that the reactor was in no condition to have been brought to power. Just as significantly, it is obvious that the owners of Three Mile Island simply were not maintaining proper standards.

At Three Mile Island there was a failure of a feedwater pump that supplied water to the steam generators associated with this PWR. This relatively commonplace incident burgeoned into a full-fledged and potentially disastrous accident because of a series of operational and mechanical failures that followed. These failures ranged from a valve that stuck open (although the control panel read closed) to the inexplicable actions of the operators, which included turning off the emergency cooling systems.

Many hours later a stable condition was reached and the core was adequately cooled. By then, however, extensive fuel damage had occurred during periods when the core was not covered with water. Because the fuel cladding overheated, a great deal of radioactivity was released to the cooling water and soon thereafter into the containment vessel. However, the containment vessel kept almost all of this radioactivity on the site, and little was released into the atmosphere. (See Fig. 3–2.)

Investigation of the accident and its causes by the President's Commission on the Accident at Three Mile Island[5] and others has indicated what was suspected by many critics of nuclear power, including some who favor its use: enormous amounts of time and money have been devoted to questions of safety, but those who build, own, and regulate nuclear power, such as the reactor manufacturers and the NRC, simply had not focused their attention properly. Procedures for maintenance and training had not been taken seriously enough. In fact a few of the utilities owning nuclear power plants appeared to be devoting insufficient effort and resources to managing their plants safely. The owners of Three Mile Island may have been one of these.

TMI also demonstrated that the utilities need more comprehensive operator training, and in some cases better control room design. The operators of the TMI reactor, largely because of their limited training, virtually ignored the fairly obvious significance of temperature and pressure measurements in the core and coolant. They had no idea that the core was being destroyed.

This led to a type of accident that, except in a general sense, was not within the scope of the WASH–1400 analysis. WASH–1400 allowed for equipment failure and operator error, but not for the extended indecisiveness of those in the control room at Three Mile Island. The ECCS pumps went on automatically, but were turned off and on at times during the accident, as were the main coolant pumps. As a result, instead of a small LOCA, which the system could have handled, or even a large LOCA, which it probably also could have handled, the operators had a growing disaster on their hands, and they were ill-prepared to manage it.

Nonetheless the amount of radioactivity released by the accident was thousands of times smaller than that from a major accident. No member of the public received a dose larger than he or she would have received from natural sources in 1979. The total number of latent cancer deaths from the accident is expected to be zero or one, and certainly no more than several. Psychological stress among the surrounding population was far more significant than the radiation effects. So the primary significance of Three Mile Island is not what did happen, but what *could* have happened.

What did happen was an accident that WASH–1400 indicated should occur about once in 400 years of reactor operation, according to the report of the President's Commission. In a sense the incident could be said to confirm WASH–1400 conclusions, but the extent of damage to the plant itself was unanticipated and constitutes a substantial financial incentive for nuclear manufacturers and utilities to improve their design, construction, and management procedures. In fact the investigations of Three Mile Island have shown that the primary causes of the accident were not design errors but rather failures in plant maintenance and operation. In the future, operating staff must be fully trained so that all available information can be assessed appropriately in order to prevent accidents from developing.

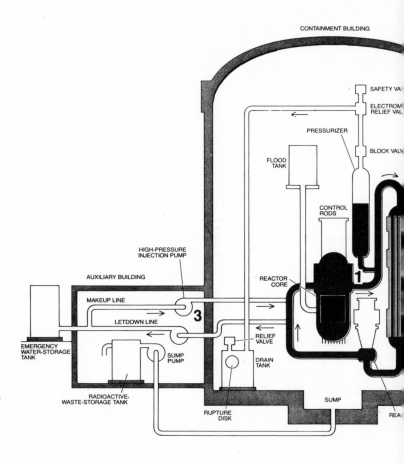

Fig. 3–2 What Went Wrong at Three Mile Island. This simplified diagram of the Three Mile Island Unit 2 reactor indicates the location of (1) the primary loop (black shading), which carries the cooling water from the core to the steam generator and back; (2) the secondary loop (gray shading) to the right of the core, which carries the water and heat away from the steam generators to the turbines; and (3) the ECCS to the left of the core, which is available to supply emergency water to the reactor core in the event of a loss-of-coolant accident. The accident at TMI began when a feedwater pump stopped operating, probably as a result of maintenance on a related system. Seconds after the rather routine loss of the feedwater pump, the resulting modest increase in reactor coolant pressure caused the reactor to scram, as designed. But backup feedwater pumps were not connected to the system as they should have been. As a result the steam generators soon had no water to boil, which caused the coolant in the reactor to heat up from fission product decay. The safety valve (top of diagram) blew open to relieve the initial pressure rise, but failed to close after the reactor scrammed. As a result water continued to pour slowly out of the reactor system for the next two hours, unknown to those in the control room, and onto the floor of the containment building, where the sump (bottom of diagram) pumped the radioactive water into the auxiliary building (left side of diagram). The radioactive waste storage tanks soon overflowed, and

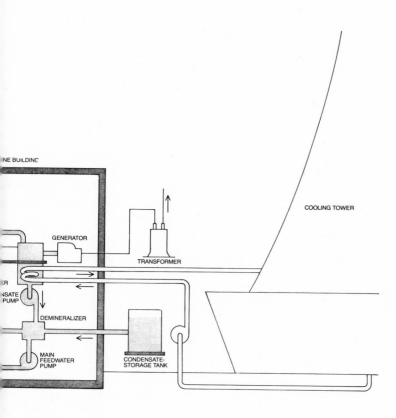

INE BUILDINC

GENERATOR

TRANSFORMER

COOLING TOWER

ER

NSATE
PUMP

DEMINERALIZER

MAIN
FEEDWATER
PUMP

CONDENSATE-
STORAGE TANK

radioactive gases began to leak over Harrisburg. Meanwhile the accident was compounded when operators, relying primarily on instruments erroneously indicating too high a water level, turned off the ECCS pumps a few minutes after they had begun working. Feedwater to the steam generators was restored when feedwater valves were discovered to be closed, contrary to procedures. But between one and two hours into the accident, operators—still not realizing how much water had been lost—turned off all the coolant pumps again, anticipating that water would circulate naturally. This did not occur, and over the next several hours the top part of the core was severely damaged. During the same period the zirconium in the cladding reacted with hot steam to release hydrogen, a highly flammable, even explosive, gas if sufficient oxygen is available. This gave rise to the famous "hydrogen bubble" in the top of the reactor vessel and, it is thought, to a small explosion in the containment building. However, it was soon known that the reactor could not have enough oxygen inside to cause a hydrogen explosion in the vessel. Although the accident was over in ten hours, the clean-up of the core itself, the walls of the reactor, and the 700,000 gallons of radioactive water still left in the containment building, is expected to keep the reactor out of service at least until 1989. [From "The Safety of Fission Reactors" by Harold Lewis. Copyright © March 1980 by Scientific American, Inc. All rights reserved.]

Of course, nuclear power must be used with care. The people in the control room *must* have a good idea of what is happening and what might be expected of them. Contrary to antinuclear criticism, perfection is not required. The system is designed to accommodate moderate failure, and risk assessment methods confirm this.

The system is also built on the premise that full attention will be given to maintenance and training. TMI disappointed many in their expectations of this kind of attention. Most other nuclear power plants are superior, but the difficulties at Three Mile Island alone indicate a failure in regulatory methods. For this reason the President's Commission recommended structural changes in the NRC and in the manner in which it views safety. Many, but not all, of these recommendations are being implemented. Certainly TMI dispelled an air of complacency and highlighted the need to devote vigorous efforts to safety analysis and design, and especially to the mundane task of checking that equipment and operators meet required standards.

The accident also caused the industry and the utilities to wake up. Most utilities cannot sustain a billion-dollar loss, and the nuclear industry cannot sell reactors in the face of such a fear. As a result, the industry and the utilities have set up, independently of the NRC, three interlinked entities. One is a Nuclear Safety Analysis Center, staffed with specialists to analyze current safety design and incidents or accidents at operating nuclear plants. A second is an Institute of Nuclear Power Operations (INPO) in Atlanta, Georgia, which evaluates operating plants and their management to see if they meet standards. A third is an insurance pool that will guarantee utilities against severe accident loss, but only if they meet standards of the operations institute. The analysis center is a theoretical resource center, and the other two help assure that nuclear plants are being operated as they ought to be. With their incentive to avoid the financial loss associated with accidents, these three groups can also make the task of the NRC easier.

The problem of human error can be significantly reduced, though never completely eliminated, by several improvements. One would be to train an elite corps of plant operators in a rigorous program much the way airline pilots are trained, who receive $100,000 per year just to supervise takeoffs and landings. Plant operators, too, would be highly trained and highly paid to handle rare emergencies and the boredom between them.

Where does nuclear power go from here? Given what we have learned recently and the changes being made, nuclear power plants now operating should continue to do so, and those under construction ought to be completed. How many additional nuclear power plants should be built can only be answered in the context of a comprehensive energy plan. Frankly, we should all cultivate an interest in using energy more efficiently in our homes, cars, and workplaces. But needs for electrical energy will persist and, to some degree, increase. For the next few decades we shall have to use coal and uranium more and more as fuels to supply electricity, and nuclear plants are likely to pose less health risk than coal plants.

WASH–1400 and the investigations since Three Mile Island, with all their uncerainties and complexities, give new assurance that nuclear power plants can be among our safest energy alternatives. Not only can multiple layers of safety systems prevent ordinary and unusual failures from endangering the public, but the improvements in regulation, maintenance, and training can prevent accidents that damage the plant itself. Continued reactor safety analyses, including investigation of mistakes at operating plants, can only make nuclear plants safer and more reliable. This and consideration of our other alternatives for generating electricity leads me to conclude that nuclear power must be considered one of our principal energy options for the foreseeable future. It is, indeed, safe enough.

Second Thoughts
JAN BEYEA

Jan Beyea is a staff physicist with the National Audubon Society in New York City. While he was a research associate at Princeton University's Center for Environmental Studies, Dr. Beyea and Dr. Frank von Hippel conducted computer simulations of the dispersion of radioactive fission products from a meltdown, which have been used extensively in licensing hearings throughout Europe and the United States. Dr. Beyea argues that reactors are inherently flawed because safety devices were always added as an afterthought, rather than being a fundamental consideration of the reactor design. He suggests that we either address these safety concerns or seriously rethink our commitment to a nuclear America.

Under normal conditions the amount of radioactivity that leaks from the core of a nuclear reactor is minute. It would appear that the health effects of a properly operated nuclear plant are minor when compared with the respiratory illnesses produced by pollutants emitted from coal- or oil-fired plants. Only gas-fired electricity stations, which routinely emit small amounts of sulfur, are as safe as nuclear power plants from a health point of view.

Proponents of the nuclear industry take this favorable comparison to be a compelling argument for nuclear power. And in fact if the radioactivity in *all* the reactors in the United States remains contained as designed, nuclear power could turn out to be a very benign source of electricity—assuming the safe handling of nuclear wastes. If a population of hundreds of reactors (thousands, if the nuclear industry has its way) can be operated for many decades without a single catastrophic accident, then the assurances made to the public by the nuclear community and the Nuclear Regulatory Commission will have been correct. The community of nuclear critics, myself included, will have been wrong (and alarmist) about reactor safety.

On the other hand, should just *one* major accident occur over the next thirty years—an accident where millions of curies of cesium-137 and other radioactive materials were released as a vapor into the atmosphere—then the total damage to human health and the environment from this one event would make nuclear power a questionable bargain. If such an accident occurs, then all the assurances of the nuclear engineers and of the congressional committees which have promoted nuclear power will turn out to have been lies. The resulting radioactive contamination would remain for at least a century as a historical monument to technological hubris, a reminder of the folly of blindly accepting the advice of a narrow group of experts. As we shall see, reactor safety is an "all or nothing" proposition; there will be no second chance.

Accident sequences which could lead to large releases of radioactivity have long been known to the nuclear industry, but were dismissed on the basis of their low probability. The strategy by which the industry and the NRC intended to keep the probability of catastrophic accidents low was based on the concept of "defense in depth." By "defense in depth" the industry meant adding enough independent barriers and safety systems to the basic naval reactor design so that the inevitable failures which occur in industrial equipment would not lead to a disaster. The concept of defense in depth is a good one in principle, but is it enough to prevent all catastrophic accidents?

The answer depends on whom you ask. Scientists in the nuclear industry judge defense in depth by the successes of the program; scientists on the other side judge it by the failures. For instance, the industry points to the fact that most of the radioactivity at TMI was contained, while nuclear critics point to the fact that only the containment building, the last defense in the defense-in-depth chain, remained between the public and millions of curies of radioactive strontium and cesium.

Both sides appear to have good qualitative arguments. If the nuclear industry is correct we may be passing up a safe energy source if we downgrade nuclear power. However, if the nuclear critics are right we may find ourselves dependent on an energy source that we will eventually curse.

The Meltdown

Although I have been studying hypothetical accidents for many years, I must admit that it is difficult for scientists to use these hypothetical models without being able to test the models against a great number of real events, which simply do not exist. The only release of radioactivity comparable to a meltdown occurred in the Ural Mountains of the Soviet Union in the winter of 1957–1958,[6] but details of this accident have been kept secret by the Soviet government (see "Accidents Will Happen").

However, I shall try to outline as dispassionately as possible the way a serious accident might unfold, without overstating or understating the results. It is not my purpose to present the "worst case," or to cast reactor accidents in the worst possible light. The consequences which I have calculated are for typical weather conditions using computer techniques generally recognized as reasonable by all sides of the

Although the accident at Three Mile Island was perhaps the most publicized nuclear accident, other more severe nuclear accidents have taken place outside the United States:

In the winter of 1957–1958, apparently a major accident involving stored radioactive material took place at an atomic military facility in a small rural town forty-three miles from the industrial city of Chelyabinsk in the Ural Mountains. Hearsay eyewitness accounts report that hundreds of people suffered from radiation burns and possible deaths, and that a large area of land was possibly contaminated. However, no solid scientific evidence is available, though several educated estimates can be found in the scientific literature.

In October 1957 a fire inside the uranium core of a carbon dioxide–cooled reactor in Windscale, England, caused 20,000 curies of iodine-131 and 600 curies of cesium-137 to escape into the atmosphere (by comparison, perhaps only 12–16 curies of iodine-131 and no cesium-137 or strontium-90 escaped out of TMI). Because of the danger posed by iodine-131 (which concentrates in the thyroid gland), British authorities quarantined and slaughtered cattle within a 200-square-mile area, removing their thyroid glands, and dumped contaminated milk into the Irish Sea. Scientists tracking the radioactive cloud found that it passed over London, 300 miles from Windscale, and eventually passed over the English Channel and over several European countries before totally dissipating. Unfortunately, the accident took place twenty-five years ago, before our increased awareness of the effects of low-level radiation, and not much useful epidemiological data was compiled from the accident.

debate.[7] For example, health effects have been calculated using numbers taken from Environmental Protection Agency estimates.

Although there are considerable gaps in our understanding of meltdowns, let us describe several scenarios which are generally accepted by many pro- and antinuclear scientists:

The reactor core overheats because of a relatively minor failure in an aging subsystem. If the ECCS fails or is turned off, large pressures build up in the reactor vessel and primary cooling pipes, causing a rupture somewhere in a weakened or embrittled* part of the reactor plumbing system. The water drains out of the core, some of the fuel rods are uncovered, and the heat from the radioactive decay pushes temperatures beyond the melting point of uranium. An uncontrolled meltdown may result.

Embrittlement in metals is caused by long-term exposure to intense radiation. Rapid changes in temperature can cause embrittled metal to fail.

Meanwhile dangerous radioactive elements boil off into a vapor. At Three Mile Island these fission products remained dissolved in the cooling water and little radiation actually escaped into the environment. However, because of human error these vapors could escape into the atmosphere through valves in the containment building left open by mistake. In such a case the four-foot-thick walls of reinforced concrete would prove useless.

Another scenario might be that the pressure of steam and other gases generated by the heat of the meltdown ruptures the containment building, either by breaking the seals around some of the many openings in the containment building's walls or possibly by forcing a crack in the concrete structure itself.

Most pressurized-water reactor domes, when properly sealed, can handle between 60 and 100 pounds per square inch (psi) of pressure (your car tire will handle 25 psi). However, NRC computer studies show that the buildup of hot gases in an accident could easily exceed 100 psi. At Three Mile Island, where containment was in little danger of being breached, some of the hydrogen gas ignited in the containment buildings, causing a 28-psi explosion. Yet today the industry is building light-water reactors (the Westinghouse ice condenser and General Electric Mark III reactors) which can only handle 10–15 psi. Even the NRC's Rogovin report expressed skepticism over whether these new reactors could handle a hydrogen gas explosion such as that at TMI. Similarly, boiling-water reactors (about one-third of all reactors) are able to handle even less pressure than the PWRs and are expected to fail in 90 percent of all meltdowns, according to the Rasmussen report.

Most studies show that the buildup of hot gases may take several hours, which might allow operators enough time to find ways to reduce the gas pressure. However a *steam explosion* could breach the containment building with little warning. Tests have shown that a reactor dome could handle an explosion equivalent to the force of one-half ton of TNT. But theoretically enough energy is stored in the molten core to cause a steam explosion equivalent to twenty tons of TNT. Steam explosions conducted in small test reactors at Idaho Falls have completely destroyed the cores. Borax I in 1954 and Spert I in 1962 were deliberately blown up with steam explosions and the SL–1 blew up accidentally in 1961, killing three workers.* Still, no large scale tests of steam explosions have ever been conducted. From experience in the metals industry we know that six died and forty-six were injured in 1958 when wet scrap metal was being loaded into a hot furnace at a Reynolds Aluminum plant in Illinois. The resulting explosion "rocked a 25-mile area," according to WASH–1400. But large uncertainties remain because uranium steam explosions and aluminum steam explosions may differ markedly in their properties.† In any case even a burst of steam that is relatively minor (compared to a steam explosion) could open the containment at certain weak points.

---

*This was a military accident, not commercial.

†Some engineers claim that molten uranium will dribble harmlessly into water. Others claim that a single "coherent" steam explosion may rupture the dome. Quite frankly, no one knows for sure.

Once the containment building is breached large quantities of steam bearing radio-active particles would rise high into the sky. The steam cloud would bend in the direction of the wind, creating a triangular plume that would spread downwind of the reactor (see Fig. 3–3). Perhaps some gray smoke, composed of tiny aerosol particles, would be visible, but even that would fade from view while close to the plant.

One hopes that off-site emergency personnel would have been notified long before the release in order to evacuate people living within ten miles downwind of the reactor. Ideally the evacuation would begin as a precautionary measure, even if operators were reasonably sure they could correct the problem. Any accident is sure to be accompanied by considerable confusion: phone lines will be jammed, preventing civil defense officials from coordinating the evacuation. (At TMI, NRC officials were often unable to get through to the reactor operators.) Roads will be tied up for miles. Panic, looting, and a general breakdown of authority might take place. Evacuation might be virtually impossible if the reactors were located near heavily populated

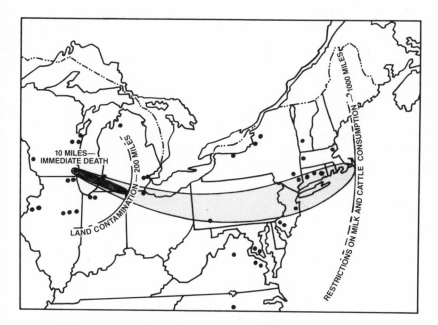

Fig. 3–3 The Environmental Effects of a Meltdown. Fission products traveling downwind in a rough 10° angle from an uncontrolled meltdown at a nuclear power plant in Wisconsin would have widespread effects on the area in their path. Our computer studies show that there will be immediate fatalities within 10 miles of the reactor, land contamination out to 200 miles, and restrictions on milk and cattle consumption out to 1000 miles. There would also be latent cancer deaths up to 1000 miles from the reactor. (We assumed, from the Rasmussen study, that about 50 percent of the volatile fission products were released in a meltdown under conditions of little rain.)

areas, such as the Indian Point reactor, twenty-five miles north of New York City, or the Zion reactor, twenty miles outside Chicago.

Assuming, however, that evacuation plans work fairly well and assuming favorable weather conditions, early deaths from radiation exposures could be kept to a minimum. My best guess is that only a few hundred people would be exposed to doses above 500 rems, which would kill them in about a month. But this guess is subject to large uncertainties. Beyond ten miles of any given plant, *no plans* exist to evacuate people, and at least a million people could be exposed to some harmful level of radiation.

The "first wave" of radioactive materials released from the reactor would spread far beyond the ten-mile evacuation radius, eventually contaminating land perhaps hundreds if not thousands of miles from the reactor. Within the first few weeks approximately 10 percent of the original radiation release would continue to be blown about the area, constituting a "second wave." This residual wave of radiation would undoubtedly cause general terror. Many people within thousands of square miles of the accident site would evacuate the area.

Those exposed to 100–500 rems would soon experience radiation sickness: nausea, vomiting. Many more people would have the same symptoms from fear alone, and would probably conclude that they had been exposed. (I doubt they would ever be convinced otherwise and would worry about it for the rest of their lives, perhaps blaming each successive illness on their possible exposure.) Thousands of cancer deaths would result years later, regardless of weather conditions or the effectiveness of evacuation within the ten-mile area.

The evacuation of the plant operators of this or a neighboring reactor poses another problem. Without constant supervision the spent fuel rods (bathed in pools of water) might overheat, causing most of the water to boil off over several days. Eventually, as temperatures of the fuel rose, the zirconium in the fuel rods would chemically react with water, causing a fire (a zirconium-water reaction) which might release even more fission products into the atmosphere. My calculations show that the resulting contamination from a damaged spent fuel pond may actually *triple* the total area contaminated.

Months after the original accident, the aftermath of this tragedy would still be felt. Restrictions would have to be placed on consumption of food from contaminated areas. If the accident occurred during the grazing season, restrictions on milk could extend 1000 miles from the accident site. Water supplies would be polluted by water-soluble fission products both from the fallout on reservoirs and the runoff from contaminated ground. Because all reactors require cooling water from rivers, lakes, and the ocean, contamination from the reactor would spread to these bodies of water, especially to communities downriver which use the water for drinking purposes.

Years after the accident, hundreds of thousands of lawsuits would still jam the courts, and damage claims would certainly exceed the Price-Anderson Act's $560 million ceiling on nuclear liability, beyond which there is no provision to pay for property damages. The government's own studies have shown that a large meltdown would result in property claims of up to $17 billion (in 1965 dollars).

Psychological stress would be another problem. The TMI accident demonstrated public concern about radiation. How much worse would the effects of a real meltdown be? It would be impossible to tell who had been irradiated. Could the nuclear reactor industry survive the anger of a million people forced to live for the rest of their lives with fear of dying from the accident? Every cancer, every stillborn birth would be attributed to the accident.

How Likely an Event?

The nuclear industry dismisses this grim scenario by stating that the likelihood of such a catastrophic accident is almost zero, and certainly within the risks accepted by the public every day. The industry also argues that the probability of such an accident can be determined beforehand by theoretical calculations. Therefore the debate hinges on the probability of hypothetical events which have never occurred in the past— events for which there is no complete historical record to estimate the risk and settle the dispute. Unfortunately, the uncertainties are so large that even a completely unbiased expert cannot accurately estimate nuclear accident probabilities.

Consider an analogy with the automobile industry. If an argument arises over automobile risks, it can be settled rather quickly by looking at accumulated accident statistics. We can project the next ten years' death toll from automobiles simply by extrapolating from past experience. There is little disagreement among safety experts as to how many deaths will take place over the next decade.

But consider how different the situation would have been in 1905 at the start of the automobile era (a situation somewhat comparable to the nuclear case today). In 1905 in order to estimate future death tolls for a mature automotive industry it would have been necessary to rely heavily on guesswork. If safety concerns had been raised before Congress, the automotive industry would probably have argued that, when operated properly, at reasonable speeds, automobiles would be quite safe. But who in 1905 could anticipate the effects of drunken driving, joy rides, the macho image of the automobile, automobile suicides, defects in manufacturing, the high density of traffic, and the high speed of modern cars? The hypothetical predictions of experts in 1905 would vary widely, just as the predictions of reactor accidents vary widely today. Anyone claiming 50,000 deaths per year would have been branded as an irresponsible alarmist.

Had an early version of Ralph Nader been able to convince the public of what was coming, automobiles and roads would never have been allowed to develop as they have, with safety as an afterthought. Initially, roads would have been built with center barriers, and passive restraints such as seatbelts and crash integrity would have been designed into automobiles.

Had we appreciated the difficulty of assuring protection from catastrophic accidents early in the history of nuclear power, had we taken a more critical look in the 1950s and 1960s at the assurances of the nuclear advocates, reactors would also have been designed differently. Reactor containments would have been designed specifically to contain the pressures and explosions that might be generated in a meltdown. Emergency cooling systems would have been designed simply (making them reliable and easy to analyze with confidence) and would form the heart of the reactor structure. Safety features would have been central to the design, not "added on." The entire program would have proceeded much more slowly. Unfortunately, the technological optimists in charge of the civilian nuclear program based their technology on naval submarine reactors and merely added on safety features for civilian use. Fortunately, it is not too late to correct this fundamental design error.

Projecting the frequency of catastrophic accidents today involves just as much guesswork as projecting automobile accident frequencies would have required in 1905. The most sophisticated computer in the world cannot compensate for the inability of engineers to visualize all possible behavior of their systems, all possible defects in design and construction, all possible operator malfunctions, all possible activities of madmen and terrorists. As a result, experts disagree about whether or not catastrophic accidents will occur over the next few decades.

The public has to learn that expert opinions are not always valid when they are not based on a large body of experience. All of us should distrust technical opinions about any new technology, whether it be nuclear or recombinant DNA technology, when those opinions are based on theory, not experience.

Certainly it would be a mistake for the public to accept uncritically the calculations of avid believers in nuclear technology. It is very difficult for avid promoters of any technology to see the bad points of their passion. Such blindness is not restricted to the nuclear engineering community. For instance, it is hard to convince solar advocates that they should consider the problems associated with certain solar technologies or to worry about pollution from wood stoves and pollution emitted from autos which burn alcohol. It is hard to convince them to worry about the increased indoor air pollution which may result from tightening buildings to make them more efficient users of energy. Avid solar proponents are sure these problems can be worked out when the time comes—just as nuclear advocates claim that every problem with nuclear power will be handled properly when the time comes, whether it be accidents, nuclear wastes, or the proliferation of nuclear weapons. It is unreasonable to expect any industry to give a fair assessment of its safety problems.

For example, instead of welcoming the recommendations made by the official commissions set up to investigate the TMI accident, the nuclear industry mounted a public relations compaign (one with little scientific merit, it turned out) to convince the public, Congress, and the regulators that less radioactivity would escape in a catastrophic accident than previously thought.

The Nuclear Regulatory Commission has not been noted for impartiality either.

The commission and its predecessor agency, the Atomic Energy Commission, were staffed with people trained in the nuclear industry. The regulators had the same optimistic bias to which the industry was prey. Until Three Mile Island the Nuclear Regulatory Commission treated the possibility of catastrophic accidents as a public relations problem, not a safety problem. Its studies were shaped to reassure the public, not to find and make improvements in reactors.

Since the TMI accident a number of long-overdue safety steps have been implemented. Evacuation plans are now required up to ten miles from the reactor, information about new accident sequences is now pooled among reactors, operators are trained to deal with abnormal occurrences, and new monitoring equipment has been added. Perhaps most important, a new attitude has developed among many NRC staff to the effect that meltdown accidents are a serious possibility. But is this enough? Moreover, I question how long these new attitudes toward safety will last now that the agency has been requested to speed up the licensing process.

As a scientist who is critical of nuclear power it would be natural for me to suggest that the public can turn to the nuclear critics for an unbiased assessment of accident probabilities. However, we nuclear critics cannot be sure we are right either. It is possible that we are too pessimistic in our judgments. Our probability numbers should not be accepted uncritically any more than the industry and NRC estimates should be. In other words, it is impossible to judge any probability number accurately. Decisions about nuclear power will continue to be made in ignorance of the true accident probability.

Whitewash 1400

Perhaps the most obvious example of the nuclear establishment's optimistic bias can be found by examining the historical record of accident prediction. The design goal for the probability of complete failure of reactor safety systems was less than one in a million per reactor per year of operation. This number was not based on any substantial mathematical calculations, but rather on a convenient number that the industry came up with in the 1950s. This goal was *assumed* to have been achieved until 1974 when the authors of the Rasmussen report actually tried to calculate the probability of a meltdown* and came up with a number fifty times more likely (one in 20,000 reactor-years) than the one-in-a-million figure, based on a detailed analysis of certain accident modes.

But how accurate was the Rasmussen report? The occurrence of the Three Mile Island accident so early in the nuclear era suggests that the Rasmussen report itself was optimistic. According to the report's most accurate estimate, an accident as severe as TMI should not have occurred for several more decades. But the occurrence of TMI shows that Rasmussen's estimate was too optimistic by a factor of ten so far as

* This estimate was not for the worst possible meltdown, but rather for a small-size meltdown.

predicting serious accidents was concerned. Although this fact alone does not prove that the Rasmussen report risk calculations are just as optimistic for more serious accidents than TMI, it suggests that this is the case.

The Browns Ferry fire in 1975 was another crucial accident sequence that the NRC and Rasmussen report failed to anticipate. In a "post-facto" analysis, the Rasmussen study group downplayed their neglect of fires by calculating that fires of the Browns Ferry type would only increase the probability of a meltdown by 25 percent. This was a self-serving result, since a higher number would have invalidated their $3-million study. Other analyses[8] suggest the risk from a Browns Ferry–type fire was much higher.*

The fact that the Rasmussen report showed that the original one-in-a-million-reactor-years figure was too high, combined with the fact that the Rasmussen report itself is probably ten times too optimistic, leads me to believe that a serious accident is *500 times* more likely than was originally predicted by the industry. As a result, I conclude that a meltdown in the next decade or two is highly likely.† After all, how long can we escape from one close call after another? And if the first meltdown occurs in a BWR, or if PWR containment defenses are (as I suspect) weaker than Rasmussen estimated, the dismal scenario I outlined earlier will become a reality.

Proponents of nuclear power argue that the industry can correct newly discovered weak points in the system. However, at some point we have to give the nuclear designers a failing grade for bad engineering practice. I would say that designers who miss their goal for accident frequencies by more than a factor of 500 do not deserve any more chances to patch up their work.

At the present time nuclear power contributes only 3 percent of total U.S. energy and 12 percent of U.S. electricity production. The need for additional nuclear plants in the next decade will be minimal. Excess electric generating capacity now exists in most parts of the country, partly as a result of energy conservation efforts by customers over the last few years. We now have the chance to halt further construction of the present design and to send the nuclear designers back to the drawing boards. If additional nuclear power plants are to be built, let them be based on a design in which safety comes first, a system which is easy to analyze, a system which is designed specifically to contain meltdowns.

If our society is to control technology rather than let it control us, we must make choices between technologies. We cannot keep giving engineers or scientists unlimited chances to run large-scale experiments which put us all at risk. Other electricity

---

*Many other incidents have occurred after the 1975 report cast doubt on the reliability of the Rasmussen estimates. For example, on June 1980 at Browns Ferry Unit 3 water seeped into the hydraulic mechanism which drives the control rods. As a result, 40 percent of the control rods failed to scram properly into the core. Though this incident did not escalate into a major accident, engineers had believed previously that a "failure to scram" was virtually impossible.

† 1 million reactor-years ÷ 500 = 2000 reactor-years. With 72 reactors in operation and 78 more under construction, we should accumulate 2000 reactor-years of operation within this decade or the next.

sources such as coal power cause health effects comparable to those that will be caused by nuclear power on a cumulative basis. Many rational people tolerate the uncertain risk of a nuclear accident in preference to the actual number of deaths from coal facilities which occur yearly. Yet I think even such people would accept the fact that a major accident is sufficiently serious, and that the probability of occurrence sufficiently uncertain, that nuclear power cannot be perceived as a desirable technology from the perspective of safety. It certainly does not appear to be a satisfactory replacement for coal. Would it not be better to move away from both coal *and* nuclear power? First, we could lessen the use of these fuels by reducing the demand for electricity through construction of efficient appliances and industrial equipment; then we could replace existing plants with windpower facilities, expanded hydropower facilities, and photovoltaic facilities.

We should think carefully whether nuclear technology is necessary. When other problems with nuclear power are considered, such as the risk of weapons proliferation and the risks from nuclear wastes, the case against nuclear power gets stronger and stronger. We already depend on one unsatisfactory source of electricity: coal power. Do we want to lock ourselves into another one?

4

Nuclear Waste Disposal

Introduction

At the dawn of the nuclear age it was assumed that the disposal of nuclear waste would present no problem. Low-level waste—such as contaminated gloves and tools used in handling radioactive materials, as well as mill tailings left over from uranium mining operations—could be buried in shallow trenches or fenced off in unpopulated areas. High-level waste, consisting of extremely radioactive spent fuel rods that must be removed periodically from the reactor's core and isolated for tens of thousands of years, could be reprocessed to extract unused uranium and valuable plutonium, which is prized as bomb material and also as a breeder reactor fuel.

Because no technical problems were anticipated, the waste was allowed to accumulate. Today we are sitting on 70 million gallons of highly radioactive military waste that is stored temporarily at government facilities in Washington State, South Carolina, and Idaho, 69 million cubic feet of low-level commercial and military radioactive waste that is stored in shallow trenches in Washington State, South Carolina, and Kentucky, and 140 million tons of uranium mill tailings that are collecting in hill-like piles throughout the Southwest. Approximately 7000 tons of spent fuel from commercial reactors is stored on racks submerged in cooling ponds or "swimming pools" at each reactor site. Although military waste has been reprocessed in military centers since the end of World War II, of the three commercial reprocessing centers planned over a decade ago, only one, at West Valley, New York, saw completion, and it was closed down in 1972 under a cloud of controversy concerning accidental contamination at the site.

One of the greatest obstacles to permanent waste disposal has been political. With more and more countries acquiring nuclear technology, fear that access to weapons-grade plutonium would result in the proliferation of nuclear weapons led President Carter in 1977 to forbid commercial reprocessing. The Reagan administration, on the other hand, has lifted the ban on the reprocessing of spent utility fuel and plans to sell the partly built commercial reprocessing facility at Barnwell, South Carolina, with hopes to have commercial spent fuel reprocessed for use in the weapons program by the late 1980s. But even the reprocessing of commercial waste would not eliminate the permanent waste disposal problem entirely; reprocessing techniques only separate the plutonium from the fission products (cesium-137, strontium-90, iodine-131), which still must be disposed of for at least 1000 years.

Another formidable problem to permanent waste disposal is public opposition to the waste sites. Although the Reagan administration intends to select and install facilities for the permanent disposal of highly radioactive waste, the public does not appear to be convinced that a suitable disposal method exists. Most citizens do not want high- or low-level radioactive waste shipped through their streets or stored in their backyards. Nine states have banned the burial of radioactive waste within their borders. Four states have declared a moratorium on new reactors until a satisfactory waste technique has been established.

The history of waste disposal in the military program presents yet another problem. Poorly designed containers built during World War II to house military waste temporarily have not withstood the test of time; 500,000 gallons of high-level waste

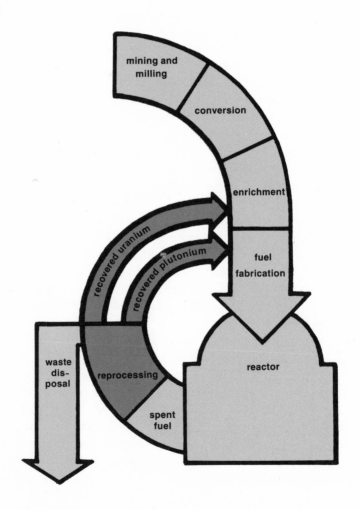

stored at the Hanford site in Washington State have leaked into the ground, and reports continue today that some of the Hiroshima atomic bomb waste from the manufacturing of the A-bomb is leaking at Love Canal near Niagara Falls and in upstate New York. These and other well-publicized bunglings in the military program have often been misconstrued as industry mistakes and have diminished the public's confidence in the nuclear industry's ability to dispose of commercial nuclear waste safely.

The nuclear industry maintains that the government is responsible for the bottleneck in the system. After World War II the government promised to take care of commercial nuclear waste disposal, as well as atomic bomb waste, in an effort to encourage the commercial development of nuclear power. Thirty-five years later the nuclear industry is appalled that the government is just now making a commitment toward a demonstration waste site.

Meanwhile critics of nuclear power are equally appalled that the nuclear industry would proceed with commercial development of nuclear power without the existence of a permanent federal repository for high-level waste. They accuse the industry of passing the buck when it blames the government for the stalemate, and argue that in

Fig. 4–1 Nuclear Fuel Cycle. (1) *Mining and milling.* Uranium ore is mined throughout the American West and Southwest, then crushed, ground, and chemically processed in order to extract and concentrate the uranium in solid "yellowcake" form, so named because of its distinctive coloration. A ton of ore will yield four pounds or less of yellowcake. (2) *Conversion.* The yellowcake is converted by chemical processes to the gaseous compound uranium hexafluoride (UF_6). (3) *Enrichment.* At one of three licensed enrichment plants (in Oak Ridge, Tennessee; Paducah, Kentucky; and Portsmouth, Ohio) a complicated series of chemical and physical processes is used to enrich the U-235 concentration within the UF_6 gas to either commercial grade (3 percent) or weapons grade (90 percent) U-235. (4) *Fuel fabrication.* Commercial-grade uranium is compressed into fuel pellets, which are inserted into the zirconium tubes that comprise the hulls of the fuel rods. The twelve-foot rods are capped, sealed, and stacked vertically to form the reactor core. Consisting of approximately 50,000 rods, the core weighs about 100 tons. (5) *Spent fuel.* After a year's operation, a third of the core, or thirty tons, must be removed as waste. This spent fuel contains several hundred extremely radioactive fission products (cesium-137, strontium-90, and so forth) and must be cooled in large ponds of water for about a year before it can be buried. Spent fuel rods are presently stored temporarily in ponds at each reactor site. (6) *Reprocessing.* Although the spent fuel is essentially waste, the core of one reactor after one year's operation still contains approximately 200 pounds of U-235 and 500 pounds of plutonium-239, a man-made element created by the chain reaction. At a reprocessing plant the valuable U-235 and Pu-239 can be chemically separated from the waste by an extremely difficult process which involves immersing the rods in nitric acid (which dissolves the U-235 and Pu-239) and chemically separating the U-235 and Pu-239 nitrates from the other radioactive by-products and purifying the nitrates (this is called the purex process). There are no commercial reprocessing plants operating in the United States today. (7) *The recovered U-235 and Pu-239* can then be recycled in other reactors; Pu-239 is only used in breeder reactors, whereas U-235 can be used to fuel either breeder or light-water reactors.

the early atomic years both government and industry ignored the problem of waste disposal.

While critics claim that we have yet to come up with a feasible, technical solution to the waste problem, many nuclear advocates maintain that the *only* problem is political, and cite a number of reports from the American Physical Society, the National Academy of Sciences, and the Ford Foundation, that have concluded that high-level waste can be stored safely for tens of thousands of years.

Most critics would admit that the disposal of low-level waste, in principle, poses no formidable technical barriers; low-level waste is shorter lived and less radioactive than high-level waste. It is the disposal of high-level waste that has generated so much concern. Every nuclear power plant produces copious amounts of high-level waste; each year, one-third of the core, or about thirty tons of nuclear waste, must be removed from each reactor and replaced. Because no federal repository exists for high-level waste and because the government forbids the utilities to reprocess the rods, the spent fuel is accumulating at each reactor site.

High-level waste is dangerous because the spent fuel rods contain hundreds of radioactive chemicals that generate large quantities of heat and radiation. When a fuel rod is placed inside the core it is not very radioactive, and can be handled without serious harm to the operator. But this same rod, when removed a year later, will radiate about 25,000 rems per hour, enough to kill anyone standing in the vicinity, and must be isolated from the environment for thousands of years.

High-level waste can be divided into two main radioactive categories. Short-lived fission products are caused by the breakdown of uranium nuclei and are hazardous primarily for the first 1000 years of waste isolation. The long-lived transuranics (like plutonium, americium, and neptunium) are formed when the uranium nuclei absorb stray neutrons, and pose a problem for hundreds of thousands of years (see Fig. 4-2). Though the transuranics are responsible for the long-term hazards of nuclear waste, the fission products account for most of the intense radiation and heat generated by the waste in the first few hundred years. When spent fuel is reprocessed, significant amounts of the transuranics are extracted, leaving the balance of the transuranics and the short-lived fission products as waste.

After ten years the radioactivity in the spent fuel rods diminishes by a factor of several thousand. The gradual decay of the radioactivity is caused by the disintegration of the nuclei in the spent fuel. For example, cesium-137 and strontium-90 each have a half-life of roughly thirty years. (An element's half-life is the number of years required for its radioactivity to drop to half its original value—i.e., for half the atoms to disintegrate. In general, the longer the half-life, the less radioactive the substance.) After thirty years a pound of cesium will have decayed to half a pound, and after another thirty years only a quarter of a pound will be left. As a rule of thumb, multiplying the element's half-life by twenty will produce a reasonable estimate of when the radioactivity will be reduced to innocuous levels, or diminished

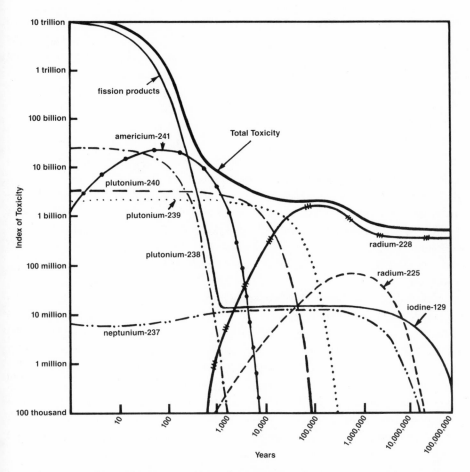

Fig. 4–2 Radioactive Waste: How Long a Danger? This graph shows the slow decrease in the toxicity (vertical scale) of the various elements in nuclear waste over thousands of years (horizontal scale). Toxicity is measured by the number of cubic meters of water required to dilute the radioactivity down to drinkable levels.

by a factor of one million. Plutonium, with a half-life of 24,000 years, would require half a million years to become innocuous.

The controversy surrounding waste disposal boils down to one question: can we trust *any* man-made structure to isolate high-level waste for tens of thousands of years given our always-changing social institutions and the vagaries of wars, revolutions, and social upheavals?

Government and industry scientists answer yes. Nuclear waste can be stored underground in geologic formations like salt domes or granite formations, which have remained remarkably stable for hundreds of millions of years—far longer than the lifetime of high-level waste. For example, because water rapidly dissolves salt, the very existence of salt formations throughout the Southwest, created when the ancient seas dried up several hundred million years ago, is proof that they have been stable against groundwater (which could carry radioactive products back into the environment) since the time of the dinosaurs. Although the government is not ruling out any options, salt formations are presently the preferred medium of geologic disposal.

To further guarantee that nuclear waste never reaches the surface of the earth, government researchers are devising multiple barriers to prevent groundwater, corrosion, heat, etc., from disturbing the waste site. First, the waste will be fused with glass, ceramic, or synthetic rock into large cylindrical blocks to seal in the toxic chemicals and thus retard the dissolution of waste in water. The French, pioneers in the glassification method, have already glassified a portion of their waste in a demonstration plant. Next the cylindrical blocks of glass will be sealed in stainless-steel canisters, and finally will be buried a third of a mile underground in a sealed geologic repository.

As formidable as these safeguards appear to be, none has progressed beyond the laboratory stage. In recent reports by the U.S. Geological Survey and the Environmental Protection Agency, geologists warned that more on-site studies must be conducted before we choose where to bury the waste. Their hesitancy is due in part to the fact that no one can predict how well the containers or the burial sites will hold up over thousands of years; the intense radiation and heat generated by these canisters could disintegrate the glass over a few hundred years and even alter the geologic stability of the rock or salt cavity. Moreover, because granite is quite brittle, the heat emanating from these canisters may cause the granite to expand and eventually fracture, creating a pathway for the escape of nuclear waste. Last, critics also point out that these high-level repositories are often located near valuable petroleum or potash deposits, which increase the chances of accidental mining long after the waste site's location has been forgotten.

Nuclear advocates contend that geologic sites can be chosen which are far from valuable mineral deposits, and that intense heat buildup in the repository can be prevented by diluting the waste in each canister or by increasing the spacing between the canisters. They admit that more research needs to be conducted into glassification and geologic formations, and accuse the critics of blocking efforts to choose or research potential geologic sites. Nuclear planners want to end the controversy by proceeding with a demonstration site—immediately.

Whether or not the waste problem is technical, or political, or both, the fact remains that a permanent disposal site in the United States has yet to be established for the tons of high-level waste that continue to accumulate. While it appears imprudent to proceed with disposal until a number of outstanding questions—namely, the behavior of waste and specific geologic media over tens of thousands of years—can be answered satisfactorily, it should also be realized that a solution to the problem will never be found unless a number of small or intermediate repositories are demonstrated in various geologic media.

Do scientists know enough to dispose safely of waste which will remain radioactive for tens of thousands of years, or will any disposal program inevitably end up as an albatross around the neck of future generations? Many Americans who otherwise support nuclear power claim that the lack of a credible waste disposal technique is, for them, the fatal flaw in the U.S. nuclear program. In this chapter a physicist and a geologist disagree as to whether, given our present technology and human institutions, scientists can successfully prevent buried radioactive waste from returning to the environment. The arguments they present should soon become the focal point of intense debate as the federal government prepares to open up a permanent waste repository in the late 1980s or early 1990s.

No Technical Barriers

FRED A. DONATH

Fred A. Donath left his position as head of the geology department at the University of Illinois in 1980 to form his own consulting firm, CGS, in Urbana, Illinois. A former professor at Columbia University, Dr. Donath is an expert on geologic disposal of nuclear waste, having performed extensive research on the mechanics of earth deformation and experimental studies of rock properties. He believes that political barriers, not technical ones, constitute the major obstacle to a successful nuclear waste program. In this essay he explains how waste can be isolated from the environment for tens to hundreds of thousands of years in a way that will pose no threat to life.

The "energy crisis" of the 1970s brought with it an unparalleled awareness of this country's energy needs for the future. It became clear that an acceptable standard of living would require the use of *all* energy alternatives and significant expansion of specific ones. For any reasonable projection this translates into considerable dependence on nuclear energy to meet the energy demands of the next two decades while new technologies (e.g., solar) and improvements in existing technologies are brought to levels that can meet future demand.

With this realization, increasing concern has developed over the growing volume of radioactive waste produced by nuclear power. Indeed critics have advocated no further expansion of nuclear energy until a satisfactory means of waste disposal has been demonstrated. This essay discusses several important questions that people want answered, such as how we can ensure that radioactive waste will be safely isolated from the environment for hundreds of thousands of years.

Various Methods of Radioactive Waste Disposal

Because spent fuel from a nuclear reactor contains a significant amount of unused fissile uranium and plutonium in addition to the unusable "waste" products, it can also be regarded as a potential energy resource. Three options exist for handling the spent fuel. The first is to dispose of it permanently; this option is commonly referred to as the "throwaway cycle." The second option is to store the spent fuel temporarily, pending a decision on whether or not reprocessing of spent fuel will be allowed. The third option is to reprocess the spent fuel and recover the unused fissile uranium and plutonium for use in other nuclear reactors. Should we choose to develop breeder reactors to help meet energy demands at the turn of the century, certainly it would seem prudent to remain flexible and view reprocessing as a viable option.

Many intriguing disposal techniques have been proposed and seriously considered, but most have been discarded as impractical, certainly before the year 2000. Ejecting waste into outer space would be enormously expensive; even the small chance of a launch mishap makes this option unacceptable. Development of the technology for transmutation, which consists of extracting the transuranics from the remainder of the waste through reprocessing and then "burning" them in commercial or breeder reactors, is not anticipated before the year 2000, if then. Burying nuclear waste in the Antarctic ice sheets raises questions about ice sheet stability. Not only might water within and beneath the ice sheets transport the waste to the biosphere, but the ice sheets themselves undergo rapid surges roughly every 10,000 years and this could even be triggered prematurely by waste-generated heat. Subseabed isolation, whereby waste is emplaced in thick sediments or rock underlying the ocean, is a future possibility, but not until more is learned about possible thermal currents and any sediment or rock behavior that could cause the waste to move back into the biosphere. Another technique would be to drill superdeep holes (as deep as 20,000 feet) and dispose of highly concentrated liquid or solid waste. Heat from the waste would initially melt the surrounding rock; later the rock would resolidify, sealing in the waste which would become an integral part of the rock structure. However, drilling such wide holes to accommodate canisters still poses practical problems, and retrieving the waste would be virtually impossible.

Even if the technology were demonstrated for these disposal techniques, they would still be unacceptable because present government policy rules that waste must be recoverable from any storage site for the first few decades. Not only must we be able to retrieve the waste in the event of leakage, but also to recover valuable unused plutonium in spent fuel, if that is the waste form. Although a few of these and related techniques show promise as future permanent disposal methods, government policy has eliminated all but one technique: burying the waste in excavated cavities in deep geologic formations such as salt beds, granite, or basalt.

Underground Cavities

Construction plans for an underground repository are impressive. A network of tunnels and storage rooms would be excavated 2000 feet underground and connected to the surface by access and ventilation shafts (see Fig. 4–3). If the spent fuel is reprocessed, the waste would be contained in solid form, packaged in corrosion-resistant canisters, and then be placed ten meters apart in holes dug into the floors of the facility. The annual waste from 400 commercial nuclear plants (the number of plants we would have in the U.S. if all our electricity were derived from nuclear) could be stored in an area no larger than half a square kilometer. In fact, thirty tons of spent fuel from a 1000-megawatt reactor operating for one year would, after reprocessing, be reduced to two cubic meters, an amount that would fit easily under a dining table. The principal reason that the waste cannot be stored so compactly, but rather would need to be stored with space between canisters, is to prevent unacceptable heat buildup. This heat is produced principally by decay of the fission products, and diminishes rapidly during the first few hundred years.

Considerable research has gone into developing containment packages that would isolate the waste successfully within the underground facility for at least 1000 years. Although the short-lived fission products such as strontium-90 and cesium-137 have an extremely large ingestion hazard* immediately after being removed from the reactor, their ingestion hazard after 1000 years actually falls below that of natural uranium ore.

Before the high-level waste would be emplaced in the repository, it would first be incorporated with some material, possibly a glass or a ceramic, with three parts of this material to one part of waste. Rugged stainless-steel canisters thirty centimeters in diameter by three meters in length which are tested for resistance to corrosion and heat would be used to mold and contain the solid waste form. The waste produced by a reactor operating for one year would fill about ten of these canisters. The canisters would be lowered into the underground facility and emplaced beneath the floor of the facility. All excavated cavities and shafts would then be refilled with the rock material removed during excavation, and sealed.

Salt deposits are a leading candidate for nuclear waste disposal for several reasons. They conduct waste heat rapidly away from the canisters because of the relatively high heat conductivity of salt. In addition, salt beds are easily mined and are believed to lack any subsurface water that could dissolve the salt. However, there are certain potential problems with the long-term stability of salt over several hundred thousand years, such as mechanical stresses caused by heat and possible changes in the groundwater regime, which have raised interest in other geologic media, such as granite. Furthermore, although salt has the advantage that it conducts heat away from the canisters approximately twice as rapidly as granite, the maximum temperature

---

*The ingestion hazard is stated in terms of the amount of water required to dilute the waste to the maximum permissible concentration for current drinking water standards.

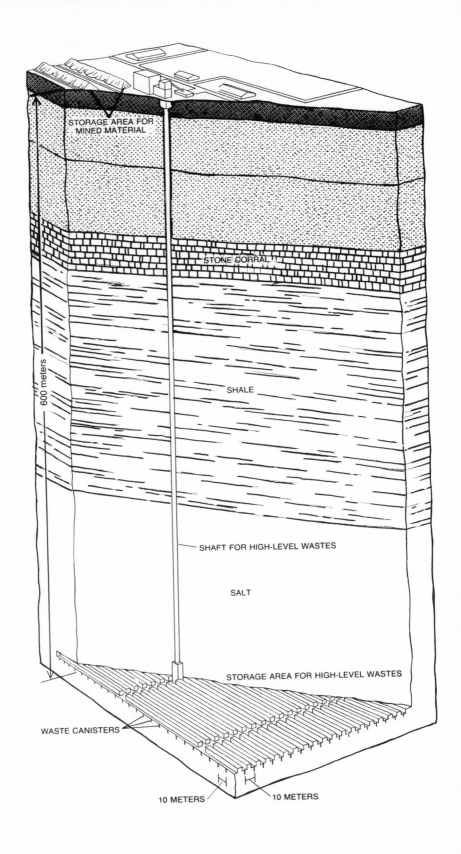

developed in granite could in fact be lower than that in salt. This is because granite has a greater capacity to store heat, which more than offsets the higher heat conductivity of salt. Moreover, a rock such as granite does not have the problems of mechanical stability or brine migration that reduce the attractiveness of bedded salt.

Currently the federal government is studying the salt beds near Carlsbad, New Mexico, hoping to establish the country's first demonstration waste disposal site by 1990. Under the Waste Isolation Pilot Plant (WIPP) project, scientists are analyzing a region of salt beds more than 200 million years old to identify some eighty square kilometers that would be acceptable for the demonstration site. Although the WIPP is designed primarily for the storage of military wastes, the government is also considering the use of twenty acres for the emplacement of 1000 commercial spent fuel assemblies, or the amount produced by one reactor operating for thirty years. Information obtained from this experiment would be used in the design of a bedded salt repository for commercial waste. Potential problems at the Carlsbad site are the subject of intensive study and discussion. The determination of the site's acceptability, as with any potential repository site, will require significant understanding of the geologic system and must include an assessment of all events and processes that can significantly influence it.

Keeping the Waste Isolated

Once a repository has been backfilled and sealed for permanent isolation, the radioactivity in the waste could return to the earth's surface only if the rock mass containing the waste were physically moved from the waste site or exhumed, or if the waste were dissolved and transported back to the biosphere by groundwater moving through cracks and pores in the rocks. Either natural processes such as erosion, volcanic eruption, and meteorite impact, or human activities such as drilling for mineral resources, could conceivably result in our exposure to the buried radioactive waste. To prevent this, stringent guidelines are needed to significantly reduce the probability that any event might breach the repository. These criteria would require that the repository be located away from known areas of volcanic activity, for example, and that the excavated cavity be sufficiently deep so that surface erosion or meteorite impact could not exhume radionuclides from the underground facility. The possibility

Fig. 4–3 Burying Nuclear Waste. Although no federal repository for high-level nuclear waste exists, government scientists favor plans to bury the spent fuel 600 meters (approximately a third of a mile) in a geologic medium, like rock salt. The nuclear waste will first be fused into water-resistant glass blocks, then sealed in stainless-steel canisters, and sunk deep into the earth. The canisters will have to be adequately spaced from each other in order to prevent large heat buildup. [From "The Disposal of Radioactive Wastes from Fission Reactors" by Bernard L. Cohen. Copyright © June 1977 by Scientific American, Inc. All rights reserved.]

of accidental human intrusion, such as drilling, could be made remote if the site were so devoid of mineral resources that it would be unattractive to future generations mining for resources.

When one first hears that nuclear waste must be isolated for thousands of years, it seems a near-impossible task for any human institution. However, we must realize that although the time scale for institutions on earth is measured in decades, the time scale for geologic processes is typically measured in millions of years. Nature has given us an example where radioactive wastes did not migrate over a period of many million years. The world's only known naturally occurring nuclear reactor is located in the republic of Gabon, West Africa. The presence of fission products in the Oklo uranium deposit provides convincing evidence that a natural reactor once operated there, nearly two billion years ago. Groundwater apparently saturated a body of uranium ore and moderated* the neutrons being generated in the ore sufficiently to create a small nuclear chain reaction. Most of the radionuclides remained immobilized in the uranium ore even though the nuclear reactor lasted for hundreds of thousands of years and created a complete distribution of fission products. This example provides good evidence that low mobility of radionuclides might be expected in nature. Of course, the conditions in actual repository sites would not be identical, and each site needs to be evaluated individually.

There appears to be general agreement that groundwater is the most likely vehicle by which waste could make its way from a sealed underground facility back to the surface. This process, however, is not an easy one. First, the canisters would have to corrode or disintegrate before the waste would leach out. Next, the waste would have to be dissolved by the groundwater. Finally, the waste would have to be transported by the groundwater back to the surface. The waste can in fact be effectively isolated if the radioactive products are transported so slowly that they would not reach the surface before decaying to innocuous levels. Because groundwater poses the greatest threat, methods have been developed and demonstrated to analyze the complex interaction between groundwater and the geologic media through which it moves. Currently, intensive research is devoted to identifying the possible influences of various events and processes on the movement of groundwater. Understanding gained by studying the generic sites can also be applied to specific proposed sites as well.

The potential hazards from nuclear waste depend primarily on three important aspects: first, how many radioactive products actually leach out of the waste and dissolve in the groundwater; second, how far and how fast the groundwater moves while traveling back to the surface; and third, how much the motion of radionuclides can be retarded by chemical interactions with the geologic media. Any one of these three can by itself control the hazards of the waste. For example, if the waste form were completely stable and inert, no radionuclides would be available to be transported by groundwater. Or if the groundwater moved very slowly and the pathways

---

*The presence of water moderates or slows down the velocity of neutrons, which increases the rate of fissioning.

for groundwater were sufficiently long, any radioactive products dissolved in the water would decay to innocuous levels before reaching the surface. Finally, if the rocks through which the groundwater moved were severely to retard the motion of the waste by chemical or other processes, the radionuclides would effectively be immobilized and therefore unable to reach the biosphere.

Clearly the choice of the waste form, whether it is glass, ceramic, or synthetic rock, will significantly influence the transport of the radionuclides. Because of this fact, great care and emphasis have been placed on choosing a waste form with long-term stability. Considerable research has been devoted to the development of waste forms that would be inert for the first 1000 years (when the fission products pose the greatest hazard) and would have low leachability after that. There is some indication that limited solubility in groundwater might even be more important than waste form leachability. In other words, even if a waste form eventually broke down completely, the limited solubility of radionuclides in the groundwater would greatly restrict the rate at which they are transported away from the disposal facility.

For those radionuclides that are dissolved and transported there is yet another important factor which considerably retards the upward movement of the waste: sorption, or ion exchange, a process by which fission products are slowed in their movement through the subsurface because the ions are trapped by the rock. Typically, sorption will reduce the average velocity of the radionuclides greatly below that of the groundwater. The Oklo phenomenon suggests that this is likely. The migration of plutonium in groundwater, for example, might be retarded by a factor of possibly 10,000–100,000 because of sorption in the rock. Scientists at several of the national laboratories, as well as independent study groups, have shown that even if groundwater were to breach a repository and dissolve radionuclides in the waste, it could take *several hundred thousand* years for these to move from a carefully selected and engineered site to the surface—long after their ingestion hazard would pose a major health problem to the public.

Radioactive Heat

Heat generated in the geologic repository by the decay of fission products poses a problem only for the first several hundred years. During this period the contents of the underground disposal facility and its surroundings will be subjected to the greatest stresses from thermal, mechanical, and chemical changes.

Critics have charged that excessive heat could open fractures in the surrounding rock and reduce the ability of the rocks to prevent the migration of groundwater. However, the buildup of heat can be controlled in several simple ways. First, the waste can be stored and aged for several years before burial, which significantly reduces the heat generated by the waste. For example, a typical canister of one-year-old waste generates about 25 kilowatts of heat. A canister of ten-year-old waste generates only about one-tenth as much, 3.4 kilowatts. The temperature of a single buried canister in typical crystalline rock could reach 1000°C for very young waste during the first year after burial, whereas the canister temperature for ten-year-old waste would be much less, about 250°C. Second, the heat generated by a waste-filled canister can be adjusted by diluting the waste concentration in each canister, for example, by increasing the ratio of glass to waste in each canister. Finally, simply by increasing the spacing between the canisters in the excavated cavity, both the canister temperature and the temperature in the surrounding rock can be lowered significantly. Increasing the spacing of ten-year-old waste canisters from five meters to ten meters would reduce the maximum canister wall temperature from about 700°C to about 400°C.

The heat generated by waste canisters in salt beds would pose another problem because it will cause small amounts of brine,* which occurs naturally in salt beds, to migrate toward the heat source. Brine occupies about 0.5 percent of the salt by weight in these beds, and was originally trapped in the salt hundreds of millions of years ago as the salt precipitated out of the early oceans. Eventually the brine will surround the canisters and corrode the canister walls, thus exposing the waste to any groundwater that might enter the underground facility. This problem can be reduced somewhat by using stainless-steel canisters that resist or retard corrosion. The brine itself would not be likely to pose a long-term threat because it does not occur in sufficient quantity to transport the waste to the surface.

No Technical Barriers

To date most models, laboratory tests, and actual experimentation have been performed to determine the *generic* characteristics of geologic media, rather than those of specific sites. Only a few of the many possibly suitable areas have yet been identified, and studies on specific sites have only just begun. Because site selection is a time-consuming and expensive activity, the public sometimes assumes, incorrectly, that little progress is being made toward a solution to the nuclear waste disposal problem.

The unfortunate experience at Lyons, Kansas, in which the engineering aspects were carefully studied but the geologic aspects were largely neglected, has not contributed to public confidence in "getting the job done." Fortunately the concerns in recent years of knowledgeable scientists and informed lay persons alike have led to extremely thorough analyses of all factors that might influence the safe isolation of nuclear wastes. The thoughtful pursuit of that objective can be expected to lead to the needed public confidence and to a successful solution to the problem.

The analysis and assessment of a site proposed for a geologic repository is never-

---

* Water saturated with salt.

theless an extremely challenging task. There is no precedent for the enormous time frames involved and the depth and complexity of the geologic information that has to be assembled and integrated with that from other disciplines such as physics, chemistry, and engineering. It is extremely difficult to ascertain the adequacy of a repository site and accurately to predict the possible ways in which waste might migrate to the surface from an underground disposal facility. The complexity of the processes involved makes it impossible to evaluate simultaneously by experiment all the factors that control transport and distribution of radionuclides through the subsurface. Unlike most technologies, nuclear waste disposal does not permit us to perform actual tests and demonstrations of the behavior of the repository system under various and changing conditions over the enormously long time period the repository is to perform. Therefore we must rely on mathematical models, with data collected over relatively short periods of time, to predict the long-term behavior of the system.

Studies at various national research laboratories and by independent research groups have demonstrated that for a carefully selected and engineered repository site, it will take many hundreds of thousands of years for groundwater to bring radioactive waste back to the surface, at which time the ingestion hazard of the water would be insignificant. Models and computer studies show that even with such severe events as earthquakes, a well-designed geologic repository can isolate nuclear waste safely.

The scientific and technological problems of nuclear waste management are soluble. Although social, political, and perhaps economic considerations will enter the picture, *there are no technical barriers* to establishing criteria that will ensure the safe and secure burial of high-level nuclear waste. However, these criteria must not be compromised by nontechnical considerations. Because no single site is likely to have all the desirable characteristics of the ideal waste repository, an optimal situation must be sought whereby if some characteristic of the site is not totally desirable, there are other factors which, taken collectively, can make the site acceptable. Certain questions undoubtedly will be resolved on the basis of social, political, or economic considerations, but it should be known that we *can* dispose of high-level radioactive wastes safely.

Will It Stay Put?

ROBERT O. POHL

Robert O. Pohl has been teaching solid-state physics at Cornell University since 1960. From 1957 to 1958 he taught at the University of Erlangen in Germany. Although most of his work is in solid-state physics, Dr. Pohl has also become known for his public warnings, in 1976, about the serious health risks posed by improperly managed uranium mill tailings throughout the Southwest. He is critical of present plans to seal nuclear waste in salt domes, and in this essay addresses some technical as well as institutional problems that plague the U.S. nuclear waste program.

In this chapter I will present a somewhat detailed discussion of a few narrow topics in the hope that this may better convey an idea of the complexity of the entire nuclear waste issue.[1] The first example will show how drastically our perception of the waste disposal issue has changed in recent years, and should thus teach us some caution with regard to our present understanding. The second example is geotechnical; it concerns the heating of the geological disposal site for high-level nuclear waste, and demonstrates some of its problems. The third example concerns uranium mill tailings, and will show the environmental risks also posed by the very large volumes of low-level radioactive wastes arising in the nuclear fuel cycle. The last example will address the least predictable of all uncertainties in the long-term management of nuclear wastes, namely future human activities, and how this issue affects the proper planning of nuclear waste disposal.

Waste Disposal Technologies—Changing Perceptions

In the early days of the nuclear age, the waste problem appears to have been entirely ignored. In a 1950 publication by the AEC, for example, radioactive wastes were considered a valuable resource, to be preserved for some unspecified later use.[2] Consequently, the high-level waste problem was nonexistent!

Seven years later this view had changed. In 1957 an ad hoc panel assembled by the National Academy of Sciences considered the disposal of high-level radioactive liquid wastes in a mined geologic disposal site. This panel found that "Disposal in cavities mined in salt beds and salt domes is suggested as the possibility promising the most practical immediate solution of the problem. Disposal could be greatly simplified if the waste could be gotten into solid form of relatively insoluble character."[3]

This philosophy of disposal was followed when the AEC proposed its first disposal site to be constructed in an abandoned salt mine in a bedded salt formation in Lyons, Kansas. The records of the 1971 congressional hearing on the AEC authorizing legislation make interesting reading.[4] They are filled with assurances by officials from the AEC and their technical experts from the Oak Ridge National Laboratories that the proposed method of disposal and its scientific and technical bases

Fig. 4–4 Abortive Attempt to Store Radioactive Waste at Lyons, Kansas. The proposed federal repository for waste at Lyons, Kansas, proved a major source of embarrassment to the AEC when it was discovered that numerous forgotten oil and gas drillings riddled the area, giving it the appearance of a piece of Swiss cheese. Not only could the holes provide a potential pathway for the radioactive fission products to reach the surface, but their very existence indicated that, in the future, mining companies looking for valuable mineral deposits might disturb the site. [Adapted from D. S. Metlay, "History and Interpretation of Radioactive Waste Management in the United States," in *Essays on Issues Relevant to the Regulation of Radioactive Waste Management,* edited by W. P. Bishop (Washington, D.C.: U.S. NRC, NUREG 0412, 1978), p. 2.]

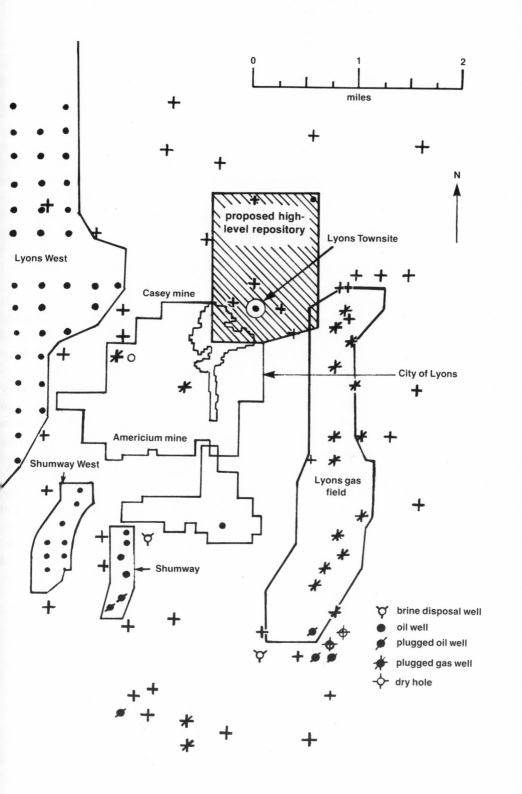

were sound. Furthermore this particular site would provide the optimum solution to the nuclear waste problem.

When the Lyons project was cancelled in 1972, two severe technical problems in particular had become apparent. The area in which the proposed disposal site was to be located was literally riddled with drill holes and cavities from numerous mining activities, as shown in Fig. 4–4. If all holes could not be found and permanently sealed—which was likely—long-term integrity of the site could not be guaranteed. The other discovery was the existence of small pockets of brine in the salt, trapped probably when the salt beds formed from drying-up oceans which had previously covered this area. These inclusions constituted, on average, only a fraction of a percent of the salt formation, but it was discovered that they migrated toward the emplaced source of heat (the simulated waste) and corroded the metal canisters. This discovery demonstrated that nuclear waste in such a salt formation would surround itself with a highly corrosive, hot liquid, which would rapidly corrode the steel canister containing the solidified waste.

As the research into improved disposal methods continued, other problems developed about the proper choice of the form in which the waste was to be buried. The most durable solid waste form considered by the AEC up to 1972 was to be achieved by fusing the waste into glass blocks, which would provide a barrier against dissolutioning (even by hot brine). In 1973, however, J. E. Mendel and I. M. Warner at the Battelle Pacific Northwest Laboratories discovered that the dissolutioning of glass was strongly temperature dependent.[5] At temperatures above 100°C (the boiling point of water) the leach rate of glass in water increased very rapidly. These measurements have since been extended to higher temperatures, and also to hot brine, and it has been shown that under these conditions glass provides essentially no barrier against dissolutioning.[6] These findings have influenced many of the current research activities. Materials scientists and geochemists are working to encapsulate the waste in more stable (i.e., less leachable) waste forms, like crystalline ceramics ("Super Calcine"[7]) or synthetic rocks ("Synroc"[8]), but none has yet progressed beyond the laboratory stage, and the debate continues about their relative merits.[9]

Also, according to the plans of the AEC in 1971 the solidified wastes were to be sealed in steel canisters which were only expected to provide protection during transportation and emplacement in the Lyons disposal site.[10] This view has also changed. Recently the NRC has proposed that waste packages should instead be designed "so that there is reasonable assurance that radionuclides will be contained for at least the first 1000 years after decommissioning."[11] I doubt whether any of the waste forms considered can be shown to satisfy this criterion which, consequently, will call for further engineered barriers, like canisters,[12] and overpacks and buffers,[13] in order to prevent a chemical attack of the waste by groundwater for at least the first 1000 years.

While rock salt had been considered since 1957 to be the prime geological disposal medium,[14] attention is now also directed toward other rocks, like granite, basalt gneiss, clay, tuff, anhydrite, etc., in order to avoid brine, which has been recognized as the "universal solvent."

While this brief historical review cannot be anything but a thumbnail sketch of the evolution of the present thinking, it does demonstrate that the method of disposal that appeared adequate to the planners of the Lyons repository less than ten years ago—glass blocks in steel canisters buried in an abandoned salt mine—has indeed numerous serious shortcomings when judged by our present knowledge. With that experience in mind, it would be imprudent to predict what the waste disposal technique may look like when it is finally accepted by the scientists and the public, and also *when* this will be.[15]

Problems Arising from Heat Dissipation

As a specific example of the geotechnical problems encountered with the disposal of high-level waste, let us consider the decay heat which is given off by the waste, and which will be trapped for very long times in the rock surrounding the waste.

Let us consider a few of the consequences of this temperature rise. The heated rock will expand. This will lift the rock above the waste, and can cause the rock to fracture. If water penetrates through the cracks in the fractured rock, it can dissolve the waste and carry it back to the biosphere. It is difficult to predict the likelihood of these events, or to determine the maximum levels of heat the repository can handle safely. As an illustrative example, we mention an intriguing point which has been raised by P. R. Dawson and J. R. Tillerson from the Sandia Laboratories,[16] and which may be important for a repository in rock salt. Because of its thermal expansion, the heated region will become lighter. Like hot air rising because it is less dense than cold air, the entire heated zone will be pushed upward. Rock salt is also plastic, i.e., it flows and can deform to fill voids, which is usually considered to be an advantage since it is self-sealing. In this particular instance, however, the higher plasticity of salt, and the fact that it becomes more plastic as it is heated, are definitely drawbacks. Dawson and Tillerson have estimated that with the kind of waste loading presently considered for waste repositories, the entire repository would rise approximately ten meters in 600 years. However, the surface would be lifted much less, and consequently some sideways motion would also take place. Such movements could rearrange the waste, making estimates of future safety of the repository extremely difficult. In reviewing this work Wendell Weart, director of the nuclear waste disposal efforts at Sandia Laboratories, declared in 1978 that "validation of these calculations will require several years of precision measurements on an experimental area."[17] Any answers derived from such studies are likely to be specific to the experimental site, and results obtained on a small scale in an experimental area may not be meaningful on the much larger scale of the entire repository. For example, the

rising of the heated rock salt depends critically on the plasticity of the rock. It is no
clear, however, how to relate the plasticity of the rock salt obtained in small-scale
experiments to the effective plasticity of a large and nonhomogeneous geologic formation.

The questions mentioned here are only a few illustrative examples of the many
unknowns which have been pointed out by geoscientists, and which have to be
answered before we can decide what constitutes a safe thermal loading, i.e., how far
the waste needs to be spread, or how long it must be aged until it may be buried. In
other words, we cannot talk about conservatively loading a repository until we know
precisely what "conservative" means. Geology as a predictive rather than a descrip-
tive science is a new endeavor, and much painstaking work still needs to be done.[18]

Uranium Mill Tailings

It is generally accepted that the spent nuclear fuel and high-level waste are very toxic
and have to be isolated permanently from the biosphere. Far less well recognized is
the fact that some low-level wastes, which are generated in vast quantities in the
nuclear fuel cycle, also present considerable hazards to future generations. As an
example we will discuss uranium mill tailings.[19]

Uranium is a relatively rare element, and only 0.1–0.2 percent by weight is contained in the ore that is currently exploited,[20] and hence several hundred thousand
tons of ore must be mined to provide the fuel for one large reactor to operate for one
year. In the milling process the ore is ground and the uranium is extracted chemically
The residue, called tailings, which contains all nonuranium isotopes of the decay
series,[21] is discarded in a tailings pond, which is eventually allowed to dry to form a
pile. In the tailings the long-lived thorium isotope Th-230 (its half-life is 80,000
years) is the source of the radium isotope Ra-226, the radon isotope Rn-222, and so
on. Radon is a chemically inert gas. Some of it escapes from the finely ground mill
tailings into the air, where it can be carried over long distances before it decays into
its radioactive daughters, polonium (Po), lead (Pb), and bismuth (Bi).

Radon gas and radium can both cause cancer, by inhalation and ingestion, respectively.[22] Removed from their geologic confinement in the ore body, the unprotected
mill tailings pose a considerable environmental hazard. On top of a tailings pile
containing an average concentration of thorium and its daughters, a person will be
exposed to a whole-body gamma-ray dose rate of 1.34 millirems per hour; this corresponds to 12 rems per year, or more than twice the maximum permissible dose for
occupational exposure (5 rems per year).[23] If a building were constructed on a tailings pile with a basement dug into it, lung dose rates of 30 rems per year would be
expected from the inhalation of the radon which diffuses through the concrete, and
from its daughters.[24] A person living in this house for approximately ten years would
be twice as likely to die from lung cancer as he otherwise would, according to the

Environmental Protection Agency.[25] His annual exposure would be twice the current limits for uranium miners (these limits are now being attacked by the National Institute for Occupational Safety and Health [NIOSH] as being too high!).[26]

While living on top of an unprotected mill tailings pile would represent an extreme case of radon exposure, even the radon carried by the wind would increase the exposure for people living in its neighborhood. According to an investigation performed for the Department of Energy, the radon-induced lung cancer rate within one mile from an existing tailings pile is expected to result in an increase of the lung cancer occurrences from *all* other sources by 14 percent.[27] Since every twentieth American today dies from lung cancer (according to the American Cancer Society),[28] a 14 percent increase of the probability of dying from lung cancer in the vicinity of the piles appears to be significant.

One might argue that these numbers demonstrate that distance apparently provides adequate protection. Unfortunately, however, mill tailings often occur as sand and hence are attractive as construction material for houses, roads, or as admixture to light clay soils. Hence, unaware of the toxicity of the tailings, people may actually seek their proximity. Numerous cases of such misuse have become known already in Grand Junction, Colorado,[29] Salt Lake City, Utah,[30] Edgemont, South Dakota,[31] Port Hope, Ontario,[32] and in the Cane Valley area of southeastern Utah and northeastern Arizona.[33] In Grand Junction, tailings sand was used extensively for over fifteen years throughout the city as construction material, until in 1966 the Colorado Department of Health found excessive radiation levels in some buildings. Current estimates are that 800 individual structures require clean-up, at a cost to the public of $16,960,000.[34] So far such remedial action has been taken on 289 individual structures, 14 schools, and 22 business/church locations.[35]

It has been estimated that by the year 2000 the mill tailings will reach 1.5 billion metric tons, ten times more than today, enough to bury the entire District of Columbia under fifteen feet of tailings.[36] How should one deal with such massive amounts of toxic waste? It appears obvious that the criterion for proper waste disposal must be a protection of future generations equal to that required for themselves by those who produce it. It follows that we have to safeguard against incidents like the one in Grand Junction where people simply did not know that the sand they used was toxic. The present practice of dumping the tailings and leaving them unprotected is therefore unacceptable.[37]

If we consider the risks posed for practically all future generations by a billion tons of tailings in hundreds of mill tailings piles distributed over many of our western states, only two disposal techniques seem adequate. One would be to bury the tailings mixed with a binder, like cement or asphalt, in deep, dry mines which would approximate the original situation of the ore body to the best possible degree. This technique might be suitable, although the availability of adequate space may be a problem, and the longevity of the binder would also be doubtful. Alternatively, chemical extraction of the thorium and radium in the milling process, and disposal in a mined geologic high-level waste repository could be chosen. This extraction technique has been demonstrated on the laboratory scale but is receiving very little attention at present.[38] The NRC has discarded both techniques as too expensive, but has reached this conclusion without including the costs of perpetual surveillance and remedial action required for its preferred disposal modes.[39]

It can only be hoped that the public insistence on a solution to the mill tailings disposal problem will be successful before the sheer enormity of the amount of waste precludes any technical solution whatsoever.

Human Intervention

The fourth and last example deals with the problem of the unpredictability of future human actions, which probably presents the most serious threat to the long-term isolation of high-level waste.

Let us inspect again the mining and drilling record of the bedded salt formation at Lyons, shown earlier in Fig. 4–4. At and near the site of the proposed high-level waste repository the salt beds are literally riddled with underground salt mines and with bore holes, shafts, and a variety of wells, giving the area the appearance of a piece of Swiss cheese. Apparently this is what must be expected if an area containing natural resources is inhabited by a technological society. It is obvious that these human intrusions will have a great impact on the long-term behavior of the salt formation in that area. Water can enter the unplugged or inadequately plugged holes, can dissolve more salt, and can cause subsidence of the overlying formations. More water can enter through the new pathways, can dissolve more salt, and finally some of the brine can return to the surface of the earth via some aquifers. The variety of scenarios is almost limitless. It was largely the recognition of the risks posed by these man-made pathways that led to the cancellation of the Lyons project, because a reliable, long-term seal of all existing holes could not be guaranteed.[40]

Researchers at the Battelle Pacific Northwest Laboratories have recently studied a worst-case scenario for the disposal of spent nuclear fuel in a salt dome,[41] shown schematically in Fig. 4–5. It was assumed that either 100 or 1000 years after the waste had been buried, unsuspecting mining companies would start solution mining of rock salt, and that the dissolved waste would be incorporated into the salt used by 15 million people for culinary purposes. This scenario would lead to a total body dose to the population over a fifty-year time span of 1.6×10^{11} and 1.3×10^9 person-rems if intrusion were to occur after 100 or 1000 years, respectively. This exposure would cause 29 million (!) or 230,000 cancer fatalities (based on the dose–health effect conversion rate of 180 cancer facilities per million person-rems, as suggested in the BEIR report).[42] Although this calculation was performed using some rather extreme assumptions, it nevertheless demonstrated that the integrity of a repository must be guaranteed for periods longer than 1000 years.[43]

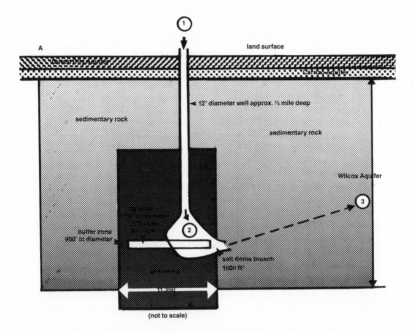

(not to scale)

Fig. 4–5 A schematic Representation of the Hainesville Salt Dome. This is one of twenty-six salt domes in the Northeast Texas salt dome basin within 200 miles of four large cities (pop. exceeding 100,000). For the worst case of human intrusion it was assumed that solution mining will destroy the repository. The scenario might be as follows: (1) water gets blown into the salt dome through a hole drilled by an unsuspecting mining company 1000 years from now; (2) the salt would be dissolved, which would cause a breach in the repository; (3) the brine could find its way to a nearby aquifer, whereby the hazardous radioactive chemicals could get into people's water systems or reach the surface.

The Many Uses for a Salt Dome

- Compressed Air Energy Storage (CAES)
- Solution Mining
- Dry Salt Mining
- Hydrocarbon Storage
- Strategic Petroleum Reserve (SPR)

- Natural Gas Storage
- Oil and Gas Development
- Nuclear and Non-nuclear Explosions
- High Energy Neutrino Astronomy
- Radioactive Waste Isolation

- Hazardous Waste Storage

Considering the many uses for a salt dome, it is not unlikely to assume that one might be disturbed many years from now.

While in principle it is possible to seal all holes that exist at the time the waste repository is filled, the real problem will be to prevent future generations from drilling new ones after the location of the repository has been forgotten. If companies drilling for minerals were to hit the repository or even a waste canister, they would soon rediscover the repository, which could then be resealed. More serious would be a near-miss *not* leading to the discovery of the repository, like the establishment of a new solution mine nearby, its operation and final abandonment, which might lead to serious consequences for the future integrity of the waste repository. Based on historic longevity of past civilizations, and considering the great instability of present human institutions, I doubt that any specific event, in particular one that is as unglamorous as waste disposal, will be remembered for more than one hundred, or at best a few hundred years. Thereafter we must suppose that our enterprising descendants would resume their drilling activities. It has recently been estimated that out of the 150 salt domes in the U.S. Gulf Coast region which would be at the proper depth to be potentially suitable for nuclear waste disposal, 95, or almost two-thirds, have undergone some form of industrial development in the roughly 100 years of industrial activity in that area.[44] Based on these facts, a human intrusion into the general area of a nuclear waste repository within a few hundred years must be considered a near certainty, if this repository is located in a geologic medium containing not only salt but any resources which humans may want to make use of (although the high water solubility of rock salt certainly compounds the problem).[45]

The proposed rule-making on geologic disposal of high-level radioactive waste published recently by the NRC contains the following sentence: "The human intrusion issue is a difficult one that is far from having been resolved."[46] I cannot see how it can be resolved at all, except by relying far more on man-made barriers of isolation rather than on the geologic medium. Work on the man-made barriers is still in its early stages, as we saw earlier. As far as the proper choice of the geologic medium is concerned, it appears that rock salt, apparently still a favorite in the U.S. and the Federal Republic of Germany, is likely to put the most stringent requirements on waste packaging.

Conclusion

The need to provide permanent isolation from the biosphere is not unique to the radioactive wastes; in fact, right now the legacy we are leaving behind in the form of chemical wastes probably represents a far more serious threat to future generations. However, if we consider the use of nuclear energy as a major source of our future energy needs, we must find adequate solutions now, rather than delaying the efforts until it is too late, as appears to be the case with many of our chemical wastes.

In discussing adequate protection from the nuclear waste, the proper yardstick, in my opinion, is not how many people will be killed by it on a statistical basis (sometimes even expressed as the number of cancer fatalities per megawatt-year of electrical energy produced). Rather, the point is whether we want to impose on future generations the need to live permanently with radiation monitors, something we do not have to do right now—apart from some unfortunate exceptions. In my opinion we should make every effort to avoid subjecting our descendants to this additional concern. This would require finding permanently safe disposal methods for *all forms* of radioactive wastes, since any disposal of long-lived radioactive species in shallow landfills, or through ocean dumping, as is currently practiced for the many forms of low-level waste arising in the nuclear fuel cycle, would be unacceptable. Properly isolating these myriads of different wastes from the biosphere might pose equal, or even greater difficulties for the nuclear industry than the disposal of the high-level waste. The disposal of the mill tailings was only one example of the large category of low-level wastes.

The solution of the technical problems involving the disposal of nuclear wastes will require a great deal of additional research, in particular where geologic processes occurring over long time periods have to be considered. Many of these questions cannot be answered through laboratory experiments and field studies alone, but also require models for extrapolation into the distant future. The validity of these models must always remain doubtful to some degree. By far the greatest challenge for the technical community, however, will be to convince a distrustful public that remaining uncertainties constitute an acceptable risk. The technologists themselves are responsible for this lack of trust, because for decades they have failed even to recognize that the permanent isolation of nuclear wastes is a major technical problem, and have consequently been unable to deliver on their promises that these wastes would be managed safely. The Lyons debacle or the mismanagement of uranium mill tailings are only two examples. Other incidents that have contributed to the erosion of the public confidence are, to name a few, the well-known leaks from the high-level waste tanks at the Hanford reservation,[47] the leaching from the poorly engineered low-level waste burial grounds in West Valley, New York,[48] and Maxey Flats, Kentucky,[49] and the catastrophic dispersal of nuclear waste in the Urals, which contaminated hundreds of square miles.[50]

We must stop belittling the technical problems of nuclear waste disposal. The only way of regaining public confidence will be through candid discussions of the problems, and through painstaking and critical research and development efforts. In addition a most meticulous clean-up of "hot spots" of radioactive waste which exist already in many parts of the country should receive the highest priority in order to give convincing evidence that the nuclear community is doing something to solve the waste problem right now. This path will be long and expensive, but it will be the only one by which the nuclear community can hope to restore its credibility, and thus hope to assure its own survival.

5

Economics

Introduction

In the 1960s when a nuclear plant cost $200 million, American utility companies regarded nuclear power as the cheapest energy source. Reactor sales continued to be strong in the early 1970s; between 1972 and 1974, 110 nuclear plants were ordered in the United States, with only 15 cancellations.

Between 1978 and 1982, however, zero orders were placed with the manufacturers, and plans for 44 nuclear plants were cancelled. One reason can be found in soaring construction costs, which have discouraged utilities from placing orders for nuclear reactors. The Washington Public Power Supply System (WPPSS) offers but one example to make utilities wary. Plagued with regulatory snarls and enormous cost overruns, WPPSS (pronounced "whoops") has not yet completed one of its three planned nuclear generating facilties (priced at $4 billion each), although construction began on the first unit more than thirteen years ago. WPPSS originally intended to build five plants, but plans to complete the other two were scrapped in 1982—after $6.8 billion had been spent—in an effort to raise enough funds to salvage the other three plants.

In addition to rising construction costs, high interest rates (up to 20 percent) have made it difficult for utility companies to raise the $1–$2 billion capital usually required for most new plants. Bewailing the tight money market, Carl Horn, Jr., chairman of Duke Power, said, "It's elementary arithmetic. You've got double-digit inflation and double-digit interest rates—and each one in itself can double the cost of a plant in 10 years." Cash-hungry utilities collectively need about $100 billion in the next five years for plant construction costs. In an effort to secure financing, utility companies have even made overtures to the Saudi Arabian Monetary Agency for loans. Explaining this decision, William McCollam of the Edison Electric Institute, a trade association for the utilities, said, "We must have access to every available capital market."

Increasingly, too, private investors are reluctant to gamble on a technology that is subject to such uncertain prospects and a fickle regulatory environment as well. Utility bond ratings, representing investor confidence, are a good indication of the fragile status of the nuclear industry. Once rated the highest at AAA, bond ratings of some of the utilities with major nuclear investments have dropped since the mid-1970s to A and even BBB.

Another reason, perhaps the primary one, why utility companies are not ordering reactors is that the U.S. is experiencing a much slower growth rate of electricity than in the 1960s and early 1970s. Until 1973 total energy sales were rising at a phenomenal annual rate of 7–8 percent. Then a series of shocks hit the economy—the Arab oil embargo, oil strikes, the accident at Three Mile Island, and sharp rises in fuel prices. By 1981 the electricity growth trend was averaging between 3 and 4 percent. Consequently recent projections for the nuclear facilities that might be needed by the end of the century are far more modest than they were ten years ago. Whereas in 1974 the federal government predicted that we would need 1090 gigawatts* of nuclear capacity by the year 2000, a recent government forecast estimated that we would need perhaps 230 gigawatts of nuclear power by the end of the century. Because facilities to support three-fifths of this capacity are already being built or planned, utility companies have little desire to order more reactors.

Although no one knows what the long-term growth rate of electricity will be, if the nation experiences a higher growth rate in the future, we could end up regretting the dearth of reactor orders today. "If we deny ourselves nuclear energy," predicted the late Philip Handler, president of the National Academy of Sciences, "we are heading for real catastrophe down the road. [There will come] a day when we will simply not have enough energy to meet the nation's needs." At a time when it would be advantageous for us to develop all domestic energy sources in order to break our dependence on imported oil, how could orders for nuclear reactors come to a virtual standstill? Those involved have divergent views as to the cause of this economic imbroglio.

Who's to Blame

The utility companies and manufacturers believe that delays brought about by the regulatory environment, specifically the NRC, have caused the industry's misfortunes. In minimally regulated countries like Taiwan or Korea, nuclear plants are ordered, built, and brought on line in seven to eight years. In the United States, although President Reagan has promised the industry a better regulatory climate, the duration is still approximately eleven to twelve years: four years to obtain local site approval and a federal construction permit (on the basis of an incomplete design) and six or seven years to build the plant. According to the Nuclear Safety Oversight Committee, a panel formed after TMI to advise the White House, issues which have no bearing on reactor safety (such as the need for power, antitrust considerations, or the availability of alternative energy sources) are all considered in the licensing process.

The industry's problems have been compounded by the critics, whose costly and

---

*One gigawatt equals 1000 megawatts, which is approximately the capacity of a typical new reactor.

time-consuming court interventions during the licensing period* have been known to delay the completion of projects for up to a year. "Lawsuits have held up orderly development of our nuclear program," fumed physicist Ralph Lapp. "In Boston, legal maneuverings by a single protestor shut down the giant Pilgrim nuclear power station for five months and cost consumers $45 million in added fuel costs." As a rule one year's delay at the construction site adds approximately $100 million to the cost of the reactor.

Although the advocates maintain that the industry's current problems are due largely to overregulation and the meddling of antinuclear activists, there are economists who argue that the industry is faring badly because the technology is an economic weakling. They point out that nuclear plants operate less efficiently or safely than the industry predicted, which costs money. It was not anticipated, for example, that the tubes inside the steam generators of the two units at Florida Power and Light would become corroded from salt water. While the units are shut down (at an estimated $51 million per unit for repairs), substitute fuels must be bought elsewhere—at an approximate cost of $800,000 per day. The accident TMI is another example of costly repairs; the capital losses from the accident (an accident which the industry describes as relatively minor) are so great—$1 billion—that the plant's owner, General Public Utilities Corporation, has argued that it could go bankrupt unless federal funds are provided to help clean up the stricken reactor.

Critics charge that the full price of nuclear power is not yet known, because utility companies cannot accurately assess what the cost of decommissioning a reactor after thirty to forty years will be, primarily because a large commercial reactor has not yet been decommissioned. Four methods are being considered: entombment (seal the reactor building in concrete), dismantlement (strip the reactor and bury the parts at low-level waste sites), mothballing (weld the door shut and guard the facility), and rebuilding on the reactor site. The cost could be anywhere from zero (if the site is used again) to $100 million, according to the Edison Electric Institute. As of 1982 these "hidden" costs were not being figured into the utilities' construction costs or rate bases.

Despite these gloomy forecasts more than seventy-eight reactors are scheduled to go on line by 1990. And the nuclear industry expects better days ahead: one major development in the 1980s has been the significant change in the political climate. The Reagan administration's willingness to encourage and assume financial responsibility for some of the costs of nuclear power seems to be a reversal of the vacillating energy policies of the 1970s. With greater assurances from the White House that nuclear power is a national commitment, Wall Street and battle-weary utilities may yet decide that nuclear energy is a solid investment. Some key federal changes include:

---

*Before a plant is allowed to go on line, citizens can appear at the licensing hearings and challenge the safety of the plant on the basis of faulty welds, the adequacy of evacuation procedures, the health effects of low-level radiation, and so forth.

○ The 1977 ban on the reprocessing of spent utility fuel has been lifted.

○ Nuclear power plant licensing procedures are being shortened to speed up the number of years required to bring a plant on line.

○ Regulations and codes that govern the operation and maintenance of the plants have been streamlined.

○ The Reagan administration has asked Congress for funds to hasten the decontamination work at the TMI reactor.

○ The breeder program at Clinch River, Tennessee, received a substantial increase in funding in the fiscal 1982 budget.

○ The government is considering the possibility of reprocessing commercial spent fuel to supply plutonium for the weapons program.

○ Energy Secretary James B. Edwards has been instructed to select and proceed with the construction of a nuclear waste repository as soon as possible.

Of course these changes have drawn fire from the critics, who are angry that a Republican administration espousing free-market principles would offer the nuclear industry a "Chrysler-like" bail-out. According to an Office of Management and Budget planning memo, funds for nuclear plants and reactors are scheduled to increase by 6 percent in 1983, to $1.7 billion. "While the President's hallowed market place is rejecting nuclear power," said Mark Hertsgaard, a fellow at the Institute for Policy Studies, an independent research group, "he himself is trying to save it. The Administration lobbied hard for $240 million to continue Tennessee's Clinch River Breeder Reactor boondoggle, allocating a whopping $500 million for long term breeder research, and greatly eased proliferation restrictions against reactor exports." Critics also argue that the federal government's policy of streamlining the licensing process amounts to little more than slashing much-needed safety regulations.

Time will tell whether the governmental policies of the 1980s can encourage utilities to build more reactors. Meanwhile the friends of nuclear power have gained a strange bedfellow—the energy crisis. As a result, many people are wondering if we can afford *not* to have additional nuclear facilities.

Do We Need Nuclear?

Nuclear power currently provides *12 percent* of the nation's electricity, or *3 percent* of the nation's total energy supplies. In terms of overall energy, the other 97 percent is produced by oil (47 percent), natural gas (26 percent), coal (19 percent), and hydroelectric power (4 percent). Critics maintain that nuclear's contribution to the nation's energy supplies is small and replaceable—3 percent is barely more than what firewood supplies. The advocates, on the other hand, emphasize nuclear's 12 percent electric contribution, which is significant and due to increase. Based solely on plants now under construction, nuclear power will supply about 25 percent of the nation's electricity by the early 1990s, according to a Department of Energy study.

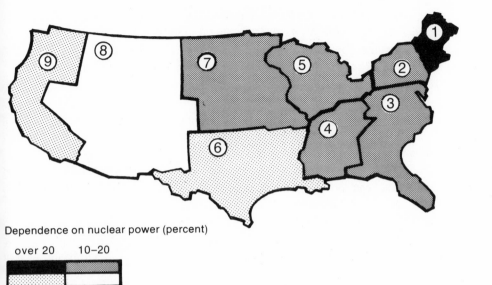

Dependence on nuclear power (percent)

over 20   10–20

1–10   less than 1

1980 Generation of Electricity by Type of Fuel (percent)

Region	Coal	Oil	Gas	Nuclear	Hydro
1	3	66	.03	26	5
2	45	27	2	15	10
3	65	14	3	16	2
4	77	2	4	11	6
5	82	4	1	13	1
6	29	1	66	3	.1
7	74	1	5	16	5
8	69	1	7	.8	22
9	3	19	12	6	58
Total for Nation	53	12	12	11	11

Note: The consumption of wood, wood waste, solar electric, and other decentralized sources which are difficult to tabulate are not included in these Department of Energy figures.

Fig. 5–1   Where Nuclear Energy Matters Most.

In the wake of the oil crisis there are many who believe that nuclear power is what will keep us from "freezing in the dark" when oil supplies eventually run out. Whether we can maintain our national security, or our standard of living, without increasing nuclear power's 3 percent energy contribution is a matter of intense debate.

To the nuclear critics the nuclear industry's tendency to equate oil savings with increased nuclear power usage is simplistic and inaccurate; nuclear power is used to produce *only* electricity, a very expensive, premium form of energy which is supplied mostly by coal: coal supplies 51 percent; gas, 15 percent; hydro, 12 percent; nuclear, 12 percent; and oil, only 10 percent (see Fig. 5–1). Nuclear expansion would displace coal plants, but not necessarily relieve us of our growing dependence on foreign oil, which has been too expensive to burn in new electric generating facilities since the 1973–1974 OPEC price hikes.

The advocates counter that while nuclear power produces only electricity, given international tensions we should develop domestic energy supplies from *all* possible sources. Decreasing nuclear power's contribution would force us ultimately to rely more heavily on coal and oil supplies to produce electricity. Although the United States, unlike other major industrial nations, has a 500-year supply of coal reserves, nuclear enthusiasts argue that coal is hazardous to mine and filthy to burn. Moreover, burning enormous supplies of coal would eliminate the pursuit of another energy option: development of a synthetic fuels industry, which is dependent on fossil fuel reserves.

Of course any argument about the need for nuclear is predicated on assumptions about the feasibility of future alternatives, such as solar energy or breeder reactors, which are addressed in Chapter 6. As to whether nuclear is cheaper than coal, the difficulty of analyzing data about relative capital and fuel costs was perhaps best expressed by Irvin C. Bupp of the Harvard Business School, who concluded that "no credible bottom line comparison can be extracted from any existing data. . . . Consciously or unconsciously, those arguing whether nuclear power is cheaper than coal are simultaneously arguing the larger issue."

Who or what is responsible for the nuclear industry's dramatic economic decline? The first two essays of this chapter discuss whether the industry's woes are perhaps the result of antinuclear obstructionism or simply the bungling and poor financial judgment of utility planners. The last two essays address a question that is central to the economic health and well-being of the United States: Can we live without nuclear power? Anti- and pro-nuclear economists clash as to whether the effects of a moratorium on nuclear plants would be negligible, or set the nation on the road to economic disaster.

The Industry's Worst Enemy

RALPH NADER AND RICHARD POLLOCK

Ralph Nader, leading consumer advocate, is the author of eleven books, including *Unsafe at Any Speed* which exposed faulty design in the General Motors Corvair, and *The Menace of Atomic Energy*. He was founder of Public Citizen, Inc. an umbrella organization for many groups working on law reform, health, safety, taxes, and energy. Richard Pollock is the former director of one of Mr. Nader's groups, the Critical Mass Energy Project, and currently the director of the Health Charities Reform Project of the Center for Science in the Public Interest, in Washington, D.C. Mr. Nader and Mr. Pollock argue that incompetence, rather than antinuclear interventions and regulatory delays, is the major reason for the nuclear industry's current financial problems.

In December 1979, as investors were eyeing potential prospects for the next year, the Washington Analysis Corporation, a subsidiary of the Wall Street brokerage firm of Bache, Halsey, Stuart & Shields, spelled out the unpleasant news about nuclear economics. "Washington Analysis Corporation estimates the overall prospects for the domestic nuclear industry as highly unfavorable," they warned clients. "Longer-term prospects continue to be clouded by severe economic uncertainties resulting from long construction lead-times, local political opposition, unfavorable utility rate structures, inability to compete in foreign markets and a failure to resolve the radio-active waste problem."[1]

Interestingly, the WAC presentation failed to cause political earthquakes or shockwaves on Wall Street. The news was received rather routinely, precisely because it reflected the views already shared by a whole range of investors, economists, and utility analysts. The WAC assessment was shaped well before demonstrators appeared at the Seabrook nuclear plant in New Hampshire or before massive citizen demonstrations occurred in Washington, D.C., New York, and San Francisco after the Three Mile Island accident.

In 1971, for instance, M. J. Whitman, an official from the U.S. Atomic Energy Commission, soberly reported to the Fourth Geneva Conference on the Peaceful Uses of Atomic Energy that "the evolution in the costs of nuclear power . . . would under normal circumstances, be classified as traumatic, rather than a successful experience."[2]

Saunders Miller, an investment banker specializing in mergers and acquisitions, is less diplomatic. To rely on nuclear energy, a projected $5.8-trillion investment by the end of the century, he said, "will constitute economic lunacy on a scale unparalleled in recorded history."[3] *Barron's,* the business journal, commented on Miller's analysis: "If the top executives of every electric utility in the nation—not to mention the utility regulators—read Saunders Miller's new book," the publication wrote in 1977, "most nuclear plant construction plans would be scrapped."[4]

Indeed the large numbers of cancellations and deferrals of nuclear power plants suggest that at least some utility executives are heeding Miller's advice. Since 1974 more than 200 nuclear power reactors that were on the drafting boards have been either cancelled, indefinitely postponed, or delayed well into the future. And not one power reactor sale has taken place in the United States since 1976. In 1978 alone, 12 reactors were cancelled and 40 delayed.[5] In the year after the Three Mile Island accident the statistics were even larger: 15 cancelled, 122 delayed.

Meanwhile construction costs for atomic energy plants continue to accelerate sharply. Reactor construction costs are increasing more rapidly than the general rate of inflation, and ten times faster than those for an oil refinery. The House Committee on Government Operations found that "nuclear plants are experiencing serious cost overruns—as much as 267 percent for one plant and more than 100 percent for others."[6]

The financial consequences for the industry have been disastrous. The U.S. Atomic Energy Commission's projected 1973 nuclear costs per kilowatt were $134. The actual 1973 costs were as much as three times that figure. In 1974 the AEC said early 1980s plant costs would be around $700 per kilowatt.[7] New plants, such as the Long Island Lighting Company's Shoreham nuclear reactor, are expected to be double that amount. In 1988 dollars, economist Charles Komanoff projects new nuclear capacity to top $3100 per kilowatt. Coal costs will hover around $1200 per kilowatt.[8]

Surprising Evidence

Naturally, understanding the reasons for these dramatic cost escalations is a key to determining the future economic prospects for nuclear power. Is the nuclear industry inherently a sick industry, or is it a sound technology wracked by government and citizen meddling?

Atomic energy executives point a finger at everyone except themselves. "Atomic energy's promise," said one electric utility executive, "is being killed by the critics, the courts, the bureaucracy, the press and the politicians."[9] Delays in nuclear construction, they argue, are one of the single most devastating factors that push nuclear costs up, and citizens and the regulators are to blame.

In flashing its allegations, the nuclear industry points to everything but its track record. Well before the first Clamshell Alliance activist set foot near New Hampshire's Seabrook nuclear plant, the federal government investigated the reasons for delay at nuclear plants. When $1 billion is riding on a plant, the government was concerned that delays can spell financial setbacks. Delays increase labor costs, postpone the day when the plant can produce revenues from the sale of electricity, and can result in the escalating price of component parts if they are to be purchased at a later date.

In 1977 the Federal Power Commission provided a partial answer to the federal government's inquiry when it assembled data from utility companies. The reasons for the delays at twenty-eight nuclear power plants due to come on line in 1974 were cited and the results were striking. The utilities, which have been known to exaggerate outside delays, told the Federal Power Commission that a total of only thirty-two plant-months of delay could be traced to safety-related regulatory changes and legal challenges from citizens. But when the delays for all twenty-eight plants were examined, legal challenges by citizen groups constituted only nine of the thirty-two plant-months of that delay, which is not much more than the five plant-months attributed to weather.[10] By contrast, 229 plant-months of delay at those same twenty-eight nuclear reactors were due to poor labor productivity, shortages, and manufacturing breakdowns.[11]

Commenting on the FPC analysis, former AEC Commissioner William O. Doub agreed that public intervention and regulation could not be blamed for the construction delays. "We all know that statistics can be very tricky, but even doubling or tripling the regulatory-related figures does not do much to close the gap [between construction and manufacturing-related problems]."[12]

Utility reports to the FPC in 1975 through 1976 showed the same pattern: 109 plant-months of delay related to equipment, 24 plant-months stemmed from regulatory safety requirements, and only 4 plant-months resulted from environmental and land-use challenges.[13]

The General Accounting Office, an arm of the U.S. Congress, agreed that most of the blame for construction delays—and cost escalation—could be laid at the doorstep of the industry itself. The GAO pointed to the largely internal reasons that significantly delayed the industry's powerplant construction times: project financing, poor utility and construction contractor management, delays in the ordering and delivery of materials, problems with labor skills, productivity and strikes, and the weather.[14]

The comptroller-general of the GAO further concluded that "prospects are not good for reducing future leadtimes for licensing and constructing nuclear power plants. In fact, GAO believes that both the [Nuclear Regulatory] Commission and industry will have difficulty in maintaining the current timeframe of 10 years."[15]

The Library of Congress added, in an analysis performed by the Congressional Research Service, that "lengthy construction times are less the result of the present Federal licensing system than they are of uncertainty about expected growth in electrical demand, and of financial difficulties of utilities in obtaining capital to fund expensive nuclear construction projects." The big bugaboo about citizen challenges and federal safety regulation contributing to delays "is largely irrelevant," the Library of Congress asserted.[16]

In fact a case might be made that while the government treats the nuclear industry with kid gloves, the public is clobbered and virtually excluded from nuclear decisions. A 1977 study by the U.S. Nuclear Regulatory Commission showed that less time is given every year to preparing an Environmental Impact Statement for a nuclear plant, although public and scientific concern about nuclear safety is increasing. In 1970 the AEC took an average of twenty-two months to complete an EIS; by 1977 the time had been cut in half, to eleven months.[17] The record suggests that, contrary to the industry's claims, it might be the citizen who has been given short shrift.

The NRC's own special inquiry group that investigated the Three Mile Island accident agreed that the public has been denied access to nuclear decision-making during licensing proceedings. In its final report the special inquiry concluded that "Insofar as the licensing process is supposed to provide a publicly accessible forum for the resolution of all safety issues relevant to the construction and operation of a nuclear plant, it is a sham."[18]

Strong words! The NRC's team went on, noting that "the vast majority of safety issues are resolved during negotiations between the NRC staff and representatives of the utility and reactor vendor which take place while the staff is performing its design review. Although the meetings that take place and the correspondence back and forth is a matter of public record, in fact the public and citizen intervenor groups seldom play any meaningful role at this stage of the process."[19]

As for the possible "meddling" by the NRC commissioners themselves, the NRC exposed the harsh reality of how limited those five men are in the nuclear licensing process: "Contrary to what the public probably perceives," the NRC special inquiry wrote, "the commissioners themselves play no role in licensing decisions except on rare occasions."[20]

Wishful Thinking

If safety regulations and citizen intervention have not caused the failings of nuclear economics, what is causing the industry to flounder? The answer is clear: the nuclear manufacturers, architect-engineering companies, and the electric utility companies. In short, the nuclear industry itself is its own worst enemy.

The Energy Project of the Harvard Business School, certainly not a bastion of nuclear critics, placed the blame squarely on the industry; they found that, among other problems, nuclear economics were based more on wishful thinking than on actual operating experience. "In order to sell a nuclear power plant to an electric utility," the Harvard specialists wrote, "its manufacturer had to demonstrate that it would produce electricity cheaper than coal or oil. The trouble was that there was no real evidence to sustain the reactor manufacturer's claim," they observed. Further, "from the mid-1960's until the mid-1970's, there was little or no effort by reactor manufacturers, by the purchasers, or by the government itself to distinguish fact from expectation on a systematic basis."[21]

Experience has been a harsh teacher for the nuclear industry. The reactor manufacturers theorized that as the reactors were built on a larger scale, the cost per kilowatt would go down. It did not: it mushroomed instead.

In 1955 a 180-megawatt reactor required more than thirty tons of structural steel and about a third of a cubic yard of concrete per megawatt. By 1965 a much larger plant of about 550 megawatts needed half as much material per kilowatt.[22] This seemingly confirmed the theory that an "economy of scale" could reduce capital costs as size increased.

But in the late 1960s, as the plants increased in physical size, the time needed to construct the reactors doubled. As the complexity increased, so did additional unexpected problems with the technology. All these factors explain why, in actual construction experience, equipment problems constitute the greatest delay.

As the plants got bigger, they should have become more efficient. Instead they became more complex and their performance declined. In tracing the actual operational data of nuclear plants, economist Charles Komanoff found that most nuclear plants performed less reliably than their manufacturers promised.

The main reactor sellers (Westinghouse, General Electric, Combustion Engineering, and Babcock and Wilcox) estimated that nuclear plants would function at high levels of efficiency, working about 80 percent of the time.* Reality exposed a grimmer truth. Often unscheduled shutdowns and unforeseen maintenance problems prevented reactors from operating at their full capacity. The Council on Economic Priorities, under Komanoff's supervision and relying solely on industry performance data, showed that in 1977 the average capacity factor for a nuclear plant was only 63.9 percent (i.e. the reactor was operating only 63.9 percent of the time).

As size increased, the capacity factor fell even further. The cumulative capacity factor for a 1150-megawatt pressurized-water reactor in 1977 was 59 percent. For General Electric boiling-water nuclear reactors the lifetime capacity factor of actual operating reactors in the 1150-megawatt range was 43–46 percent through 1977.[23]

The result? In 1976 the Bank of America circulated an internal memo questioning its financial backing of commercial nuclear power projects. A year later, in May 1977, the General Electric Company told the federal government that unless there was a turnabout in nuclear power prospects, it would eventually stop manufacturing reactors.[24]

General Atomic, another reactor vendor, has already pulled out of the commercial reactor market. Babcock and Wilcox, the manufacturer of the now crippled Three Mile Island plant, has closed the doors on its reactor factory in Mount Vernon, Indiana. The 1000 atomic power plants boastfully predicted to be in place by the year 2000 have withered and current government authorities expect about 200 to be available for operation by that time, if all goes well.

Wall Street, abuzz with jitters about nuclear power, is counseling investors to stay away from utilities with large nuclear plants. Moseley, Hallgarten, Estabrook & Weeden, a well-known Wall Street investment firm specializing in utility stocks, rated nuclear-based utilities as "unsatisfactory." Their recommendation: "We would

---

* This is referred to as the capacity factor rate. Because of the sophisticated nature of the technology, an 80 percent capacity factor is considered quite good.

expect that companies which are primarily coal-fired and operate in a reasonable regulatory atmosphere could turn in the best earning and price performance in the years ahead."[25]

Lewis Perl, of National Economic Research Associates, offers similar advice. "If I were a utility executive today," Perl told *Barron's,* "either with a nuclear plant a short distance under way or with one I hadn't started building, I'd get out of it and I'd never build another. Because I think the risks to your stockholders are just intolerable."[26]

Perhaps the nuclear industry understood the risks associated with this technology from the beginning. In the mid-1950s when commercial atomic energy was being debated in Congress, corporate leaders of the prospective nuclear firms testified that before the first stone could be put into place they would need a public limitation on financial liability and assurance that they would not go broke with the first plants.

Their demands were acknowledged: Uncle Sam told the electric utility companies that their initial reactors would be no-risk ventures. Washington would cover the companies' losses if the atomic reactors proved to be a losing proposition. These early plants were called "turnkey" plants, which meant that the government would share the costs, guarantee that the industry would sustain no substantial financial losses, and "turn the key" of the plants over to the private utility companies as their own.

With the prodding of corporate lobbyists Congress included more subsidies. The legislators imposed a financial limitation on corporate liability in case a reactor caused widespread destruction. The company's total financial responsibility was established at $60 million, a token amount considering that AEC studies claimed a nuclear power disaster could wreak tens of billions of dollars in damage and injury. The U.S. government would shoulder the additional $500 million. The Price-Anderson Act still shields the nuclear industry today, providing a limitation of $560 million.

Nuclear economics have failed, and will continue to fail as the costs spiral from other problems: long-term radioactive waste disposal, decommissioning and dismantling of nuclear plants, and new safety requirements (as a result of TMI) for all existing and unfinished plants.

As the NRC's special inquiry group noted, the federal licensing system has been a sham. The presidential commission that investigated Three Mile Island agreed: "the evidence suggests that the NRC sometimes erred on the side of the industry's convenience rather than carrying out its primary mission of assuring safety."[27]

What We Can Do

The role should be reversed so that the public can participate in licensing decisions. Citizen groups that contribute to nuclear safety issues should not only help review the licensing process, but should be given funding and tools to counter the industry's well-financed presentation. As the AEC's Atomic Safety and Licensing Appeal Board noted, "Public participation in licensing proceedings not only 'can provide valuable assistance to the adjudicatory process,' but on frequent occasions demonstrably has done so."[28]

Even James R. Schlesinger, who served as the chairman of the AEC, testified on the importance of public participants in nuclear proceedings. "We had quality assurance problems at [New York's] Indian Point [reactors]. The intervenors picked those up. We say 'all power to them,' if there are problems let them be heard."[29]

The NRC's special inquiry into Three Mile Island agreed with the Schlesinger assessment, saying, "intervenors *have* made an important impact on safety in some instances—sometimes as a catalyst in the prehearing stage of proceedings, sometimes by forcing more thorough review of an issue or improved review procedures on a reluctant agency." "More important," the NRC staff declared, "the promotion of *effective* citizen participation is a necessary goal of the regulatory system, appropriately demanded by the public" (emphasis in the original).[30]

In 1979 the NRC special inquiry group came out in favor of funding for citizen groups involved in crucial nuclear licensing and rule-making proceedings. As they correctly noted, "the problem of providing for increased public involvement in the decision-making process cannot be separated from the question of providing public funding for such activity." They stated that since these groups contribute to public safety, "they should be reimbursed for their expense." And the special NRC group noted, "Other agencies have programs to fund citizen participation and even, as under the Clean Air Act and Federal Water Act, citizen lawsuits."[31] The citizen funding for lawsuits would be awarded only after the suits and if the citizen side wins.

For years the atomic energy industry has tried to sell the public a false world of technological and economic paradise, where, as former AEC official Lewis Strauss once predicted, nuclear electricity would be "too cheap to meter." But consumers live in a world where the harsh reality is that nuclear economics are indisputably falling, and getting worse every day. The expensive plants drain the available capital that could be used for other safe energy alternatives like solar power.

Certainly the solar promise could be quashed by the insatiable capital appetite of the nuclear industry. That would be the ultimate irony, given the 1979 U.S. Department of Energy study which showed that harnessed renewable energy resources could make California, our most energy-intensive state, fully energy self-sufficient by the year 2025 while doubling its population and tripling its state economy.[32]

Ultimately, as the full $7-billion (or more) repair bill of Three Mile Island is realized,[33] nuclear power economics will appear less and less attractive. Even without another nuclear catastrophe, it is clear that atomic energy is too high a price for our electric bill. If an accident does not end nuclear power, its economics will. For decades, although the industry has tried to ignore its record, Wall Street has not, nor should consumers.

On the Road to Recovery

TONY VELOCCI, JR.

Tony Velocci, Jr. is a senior editor at *Nation's Business* magazine, a leading business journal based in Washington, D.C. Mr. Velocci has written about defense and energy-related topics for a variety of publications and lectured extensively on the future of nuclear energy. He believes that the financial problems of the nuclear and utility industries are intertwined, and those difficulties stem primarily from state regulatory policies and obstruction of antinuclear activists. In this essay he outlines what is needed in order to direct private investors back to nuclear power.

America's conventional nuclear power program is at a crossroads, and the final direction it takes will help shape this country's future in the closing decades of this century and much of the twenty-first. While virtually every other industrialized nation vigorously expands its use of nuclear energy, the pioneer and current leader in this technology—the United States—contemplates phasing it out.

Among the factors that will eventually determine the future of this country's commercial nuclear power program is the domestic political climate. Witness, for instance, how the four-year absence of a clear-cut nuclear policy during the Carter administration helped slow nuclear power's progress to a snail's pace. Economics and public attitudes will also play critical roles, as will the pressing issues of plant safety and waste disposal.

But there remains another aspect of the nuclear question that is likely to be just as decisive: the utility industry will determine in large part whether this country has a viable nuclear program in years to come, or whether nuclear has only a limited future.

The problems facing the utilities' development of nuclear energy are many, and among the most formidable are regulatory delays and disruptive antinuclear protests which have helped to multiply the cost of building new plants. Both problems threaten to render nuclear an unacceptable financial risk for utilities already weakened by outmoded regulations, nearly a decade of vacillating national energy policy, and the effects of double-digit inflation.

There is no question that nuclear power faces a barrage of serious problems, not the least of which is the utilities' ability to finance and build atomic generating stations. But don't write off nuclear yet. First of all, President Reagan's nuclear policy has been crucial. It means that the seventy-two existing reactors will probably be allowed to continue producing base-load electricity and work on the seventy-eight under construction will likely continue toward completion.

Carl Walske, president of the Atomic Industrial Forum, the industry's trade group, believes the country's nuclear electric generating capacity will nearly triple in the next ten years, and by 1990 it will account for 16 percent of the U.S. electric power

capacity, compared to 9.2 percent in late 1980. In terms of actual electrical generation, the AIF expects nuclear power to double its role in the 1980s, from 11.4 percent
in 1979 to a projected 22 percent in 1990.

One reason for this optimism was the nuclear industry's swift action following the
Three Mile Island accident to upgrade overall operations, particularly those involving
safety. Also, periodic polls leading up to the presidential elections reflected an
increasing awareness and sophistication among Americans concerning our energy
predicament. (A Harris poll at the end of 1980 found people evenly split on nuclear
power; 47 percent were for it and 47 percent were against it, with the remainder
undecided.)[34]

For the present, nobody is ready to issue nuclear power a clean bill of health in
this country, but there are a few knowledgeable people today who would argue that
the prognosis for possible recovery—even a healthy life—has decidedly improved.
Certainly nuclear power occupies a unique place among energy sources. In recent
years it has been the fastest growing contributor to domestic energy supplies, exceeding coal by 25 percent.[35]

There is further evidence of the viability of nuclear power. In 1979 it accounted
for at least 50 percent of total electricity production in four states; Vermont, 78
percent; Maine, 60 percent; Connecticut, 53 percent; and Nebraska, 50 percent.
Regionally, New England leads the nation with 34 percent of its electricity generation
based on nuclear energy, followed by Middle Atlantic states (about 17 percent),
South Atlantic (15 percent), and East North Central and West North Central (tied at
14 percent each; see Fig. 5–1).

The reason why nuclear power has experienced such substantial gains is basic: it
produces power safely and cheaply. If nuclear were to continue the growth trend
experienced up to 1979, by the year 2000 it could increase U.S. domestic energy
supplies by the equivalent of nearly ten million barrels of oil a day over the amount
used in 1978.[36]

Utilities are not forced to choose nuclear power. It just happens to be cheaper than
coal, its chief competitor in generating central-station electric power. But the regulatory and political atmosphere are so negative that the economic advantages of nuclear
power are gradually diminishing.

The Burden of Regulations

Clearly, utilities are not rushing to build nuclear power plants. Many utility executives, like Russell Britt, executive vice-president of Wisconsin Power and Light Co.,
say that part of the reason is confidence in the regulatory process. The consensus, at
least up until very recently, was that the Nuclear Regulatory Commission could no
longer be counted on to reach decisions about the future of nuclear power on the basis
of safety, the merits of the issue, or nuclear's past record.

Another major obstacle is that increases in utility rates granted by state public utility commissions have usually fallen short of what the utilities themselves require to offset investments in nuclear technology. Rates ideally should be made for the future, but in reality they are often based on what happened in an arbitrarily chosen period in the past, according to Leonard S. Hyman, senior industry analyst and vice-president of the securities research division of Merrill Lynch, Pierce, Fenner and Smith, Inc. Regulators generally start with an outdated test year and then take a year to reach a decision, so that rates the following year are actually based on two-year-old data. "A lot of water flows under the bridge in two years, especially in an inflationary economy," says Hyman.

Such regulatory lag is a basic reason why utilities have been unable to realize adequate returns in the past. Some regulatory agencies are now attempting to reduce or eliminate lag by basing decisions on projections.

According to financial sources, the return on equity allowed by regulators has not kept up with returns earned in other regulated industries, industry in general, or the return offered on new bonds.[37] Because of the way in which rates of return for utilities have been determined, regulations have in effect artificially depressed the cost of central-station power below its true cost to consumers.

This in turn has done two things: First, the development of competing forms of energy has been inhibited, just as oil and gas price controls by the federal government discouraged energy firms from exploring and developing new harder to find, harder to recover oil and gas deposits. Second, an artificial demand for electricity was created above what happened in the oil and gas industry. While every other nation paid the world market price for petroleum, the U.S. kept a lid on prices for years, insulating Americans from the true cost of petroleum products.

In both cases—nuclear and conventional energy forms—government regulatory policies destroyed any incentive to conserve, although energy consumption began declining in mid-1979 due in large part to increasing energy costs. For utilities this artificial demand put immense financial strain on their resources.

Another regulatory drawback is that fuel costs can generally be passed on to customers with little or no delay, but increases in capital expenditures (including those resulting from adding a new generating plant to the rate base) require regulatory approval, which is spread out over months, sometimes years. The big risk in constructing a new fossil-fuel plant is the cost of fuel. When a nuclear plant is completed, the biggest risk is recovering capital costs. It should come as no surprise that many experts, like Hyman, believe strongly that the regulatory process as it now exists is completely unsuited for a changing industry operating in an inflationary environment.

In the opinion of securities analysts, the single most critical factor determining the credit position of electric utilities is the quality of the regulatory jurisdiction in which they must operate. Eunice T. Reich, vice-president of Merrill Lynch, points out that the better regulated utilities retain a higher percentage of their earnings than those with less favorable regulation.[38]

"Regulators are just now beginning to recognize that rates have to be higher and more expedient, but the process of reaching that determination is very cumbersome," says Carl H. Seligson of Merrill Lynch White Weld Capital Markets Group. "And his situation has been going on for years."

Financial Outlook

As utilities entered the nuclear age a number of factors, operating together, depressed their profitability and the marketplace's regard for the industry. This change of fortunes will have a direct bearing on how many nuclear plants are ultimately built. "The financial situation of the industry and regulatory procedures both work against launching into the most capital-intensive projects," says Hyman. "You can't get much more capital intensive than nuclear projects."

Investor-owned utilities constitute about three-fourths of the total utility industry. Stockholders' investment in these companies has declined sharply in the past fifteen years.[39] In 1965 utility bonds yielded 4.6 percent, or almost eight percentage points less than what stockholders' money was earning in an electric company. As a result investors were so enthusiastic about the earning power of utilities that they were willing to bid below book value.

By 1970 return had declined because of inflation, higher fuel costs, environmental regulations, the delays in obtaining rate relief, and the unwillingness of regulators to face up to the need to reprice electricity upward. Interest rates were climbing. Stockholders were expected to take the greater risk of equity investment for roughly three percentage points more in return than from bonds.

The slim three-percentage-point margin narrowed to only one by 1975. Four years later stockholders were earning a return about the same as that available to bondholders. In 1979 investors were paying a price only three-fourths of book value, which meant that despite the rising cost of power plants, the average investor would not pay a price for the utility stock that was at least equal to what previous investors historically had been willing to invest in utilities.

Today earning power on money that utilities put into new plant is so poor that investors immediately mark down that investment to a value lower than the cost of the plant. (Electric utility stocks now sell approximately 20 percent below average book value.)[40] The result of this kind of trend is a tendency among utilities to minimize new investments and avoid capital-intensive projects.

Roughly half the capital raised by utilities is in the form of debt. The better the debt standing, the easier—and less costly—it is to raise additional debt. In 1965, 89 percent of the electric utilities had the two highest bond ratings. Ten years later only 7 percent had the two highest ratings. Generally, a low debt rating means having to pay more to borrow money during periods of turbulence in the capital markets such as now exists.

The Cost of Delays

Delays in completing nuclear power plant projects have been a major contributing factor to the deterioration in bond ratings. A two-year delay on a project will add approximately $200 million to the basic cost, according to the New York–based investment firm of L. F. Rothschild, Unterberg, Towin, Inc.

The Tennessee Valley Authority (TVA), which serves all or part of seven states and is the largest single electric generating system in the U.S., knows all about the costs of delayed nuclear power projects. In the mid-1960s TVA's development strategy foresaw construction of seven nuclear generating stations. During the intervening years the costs of these plants have grown from $6.8 billion to $17.6 billion. The bulk of the $10-billion cost increase represents the cost of delay.[41]

Every delay, no matter what the cause, introduces opportunities for more interventions, lawsuits, and requests to reopen old issues. ''The performance of nuclear critics has been obvious,'' charges A. David Rossin, director of research for Commonwealth Edison Company. ''They have taken advantage of delays to create more delays, and in an inflationary economy these force up costs. Cost estimates figured when commitments are made . . . include some allowances for contingencies but do not assume that lengthy and unpredicted delays are inevitable. To make such assumptions would arbitrarily inflate the costs.''

When a project is delayed, a utility must figure out how to finance expenses for which it is responsible, even on incomplete work, until the plant becomes an income-producing asset. In the case of outright cancellations—there were eight in 1980—a utility must face the difficult questions of how much money will be lost through write-offs and how much of the write-off can be passed on to ratepayers.

Uncertainties Ahead

Compounding the utilities' perplexing dilemma—to invest or not to invest in nuclear—is the added uncertainty of what the future demand for electricity will be in, say, twenty or thirty years. The utilities know that the trend now is toward conservation, but many economists and energy analysts believe this will reverse itself by the early 1990s and existing reserve margins will drop dramatically.

Utilities argue that in times of uncertainty—and we face perhaps three decades or more of it so far as our energy future is concerned—it is far better to have generating capacity on hand than to need it and not have it.

The Tennessee Valley Authority, which serves 110 municipal electric systems, forecasts a growth in electric consumption of 4.6 percent a year through 1990, compared with the historic 7 percent increase. However, even at a 4.5 percent growth in electric consumption, a doubling of generating capacity will be required in less than sixteen years, versus ten years with a 7 percent growth.[42]

Central to the utility industry's predicament is that its product—electricity—cannot be easily or economically stored. Therefore utilities must have enough plant to meet demand at all times, including the period of peak demand. Uncommitted power capacity, otherwise known as reserve margin, now runs to more than 30 percent on a national basis and will likely remain high throughout the 1980s. It is not expected to fall to 22 percent—now regarded as the optimum figure—until 1990.[43] But composite figures conceal the fact that the adequacy of reserve margin varies widely by region. Parts of California, for example, appear to be headed for a shortage of electrical capacity. So does New England. A number of other geographic areas, while now reporting adequate reserve margins, could well run short if local demand proves volatile.

Prudent energy planning should be based on the assumption that new electricity capacity through the end of the century will largely utilize existing technologies. While coal has great growth potential, it is expected also to be used for gasification, coal liquefaction, and direct burn in industry. Reports the U.S. General Accounting Office: to the extent that coal cannot meet all these demands, nuclear is the only developed domestic energy source able to provide large amounts of new electric capacity.

What direction is nuclear likely to take? It is still too soon to tell for sure; most financial and energy authorities say that signposts indicate conflicting readings. As one industry analyst puts it, dimmer than its proponents had expected, better than its opponents hoped.

The likelihood that most licensed nuclear plants will survive does not mean that nuclear power, or utilities for that matter, are out of the woods. "Low rates of return on equity, the scarcity of financing, and high interest rates will put privately held utilities under great pressure," says D. D. Danforth, vice-chairman and chief operating officer of Westinghouse Electric Corp. Irvin C. Bupp of the Harvard Business School flatly predicts that upward of a dozen utilities will fail in the mid-1980s, taken over by state and local governments.

Until electric utilities manage to develop new methods of financing and meeting the needs of the public other than at the expense of security holders, or regulators show greater responsibility in setting returns and terms of service, there seems little reason to expect the electric utility stock to become anything other than a vehicle for high current income and minimal growth in dividend.

To stimulate commitments to new nuclear plants, a mechanism will have to be devised to shorten licensing time and bolster confidence that once-issued construction permits, barring genuine safety considerations, will remain in force and schedules will be insulated from capricious interference.

Cautiously Optimistic

In the event that nuclear remains stalled and utilities do come up short in generating capacity in the next decade, what will be the backup energy source? Coal? Solar? By then oil will be totally out of the question. Says Hyman of Merrill Lynch: "I would hope that after coal strikes, the oil embargo, the Iranian revolution and several drought on the Pacific Coast, we would have developed some appreciation for the virtues of diversity of fuel supply."

In other words, it seems foolish to act as if alternative technologies are going to be available at the right price just when we need them. Past experience in attempting to predict the path of new technology bears this out.

It appears unlikely that nuclear technology will stand still, unless, that is, we simply give up on it. And the probability that that will happen is rapidly diminishing especially in view of the Reagan administration and a Republican Senate. There is no question that President Reagan is strongly energy-production oriented, and he is getting a lot of help from Congress where policy-making will be strongly affected by Republican gains. The elevation of Sen. James McClue (R., Idaho), one of Congress's most ardent nuclear supporters, to the chairmanship of the Senate Energy Committee is but one example.

In both Houses of Congress, and within each major party, there is growing support for nuclear power. Typical of the sentiment among senators and representatives is the view of Sen. Pete Domenici (R., N.M.), chairman of the Senate Budget Committee and a member of the Senate Committee on Environmental and Public Works: "I am convinced that America's crude oil dependence will not be appropriately addressed unless we make a firmer commitment to nuclear power."

Energy Secretary James B. Edwards is another advocate of nuclear energy. He served as chairman of a nuclear energy panel sponsored by the National Governor Association when he was governor of South Carolina.

Industry leaders are under no illusions, however. They know that devising politically acceptable solutions to nuclear's problems will be a formidable task indeed, and a mere change in administrations alone offers no assurances for a bright future. Nonetheless, Reagan's election did provide a psychological pill of enormous dimensions.

Most utility and financial executives expect to see no new domestic orders for nuclear plants until at least the mid-1980s, unless there is a drastic change in the demand for power.

Overall, advocates of nuclear power are cautiously optimistic about the future. The Nuclear Regulatory Commission is trying to streamline its procedures for issuing operating licenses to newly built commercial power reactors. Three new plants were issued full-power operating licenses in 1981, and the NRC's new procedures may result in operating licenses for another seventeen new plants in 1982, although the exact number remains uncertain. Developments like these tend to reinforce what more and more people are suspecting: nobody is ready to issue nuclear power a clean bill of health in this country, but the prognosis for possible recovery appears to be steadily improving. Consider for a moment the statement earlier this year by Sen. Gary Hart (D., Colo.), a moderate, at best, when it comes to nuclear: "If we can make nuclear power safe, it has a future. If we can't, it doesn't. It's that simple. . .

But we can't get off our addiction to Persian Gulf oil, which we must do immediately, without nuclear power.'' Senator Hart, you may recall, headed the congressional committee in 1979 that investigated Three Mile Island.

From the standpoint of investors, such signals are crucial. The reason? Investor anxiety levels run high or low, depending on the specific nuclear news of the day.

Nuclear power has a meaningful contribution to make to the long-term energy security of this nation. Realizing that contribution, however, is predicated on first being able to finance and build nuclear power plants. Utilities can cope with stringent regulations and even bad regulatory decisions, but they are not well-equipped to cope with indecision and interminable delay, advises Rossin of Commonwealth Edison. And he is absolutely correct.

The bottom line is that the future of nuclear power in the U.S. will be decided by what constitutes a prudent business decision for the boards of directors of utility companies. In the final analysis, advises Hyman of Merrill Lynch, it is academic whether nuclear power is a blessing or a peril if those who would build the power plants choose not to do so.

Living Without Nuclear Energy

VINCE TAYLOR

Vince Taylor is a staff economist for the Boston-based Union of Concerned Scientists, and author of several UCS publications, including *Energy: The Easy Path*. Previously Dr. Taylor was an economist and analyst for the Rand Corporation in Santa Monica, California. From 1974 to 1979 Dr. Taylor was a consultant to the Arms Control and Disarmament Agency, and to the Energy Research and Development Administration (ERDA, predecessor of the DOE) on nuclear power and its relation to the spread of weapons. In this essay Dr. Taylor describes why he believes that nuclear power is unnecessary.

'Nuclear electricity will be too cheap to meter,'' claimed its early promoters. This once-upon-a-time promise quickly fell victim to the expensive realities of trying to make a difficult, inherently dangerous technology safe and reliable. Indeed nuclear costs now appear to have risen above those for coal-generated electricity.[44] Meanwhile, however, the energy crisis has exploded, and nuclear supporters have been quick to shift the focus of their argument to take advantage of people's new fears of oil dependence.

Since Three Mile Island, television ads sponsored by the major nuclear manufacturers and electric utilities have proclaimed that nuclear power is ''needed to get that foreign oil monkey off our back''; an extended ad campaign on energy policy by the Mobil Corporation has played the theme that ''power derived from the atom is essential to solving the energy problem''; and a *Wall Street Journal* editorial has cited lagging nuclear construction programs as an example of the country's lack of determination to solve the energy crisis.

These media arguments have been temporarily effective in convincing the public and Congress that the nation must rush forward with nuclear power despite the many unresolved questions about its safety and economic desirability. But it is only a matter of time until they too are proven false by the force of reality. The argument for nuclear necessity pictures a world that does not exist, one where electric utilities consume a major proportion of all oil, have no choices except to continue to use oil or to build nuclear plants, and risk creating electrical blackouts everytime they delay or cancel a nuclear plant. Nuclear power's contribution is portrayed as large and indispensable. It is neither.

Because nuclear power receives so much attention, many people will be surprised to learn that it provided only 3.5 percent of total (primary) energy consumption in the United States in 1980.[45] Its share was small because (1) only one-tenth of electricity was generated by nuclear energy, and (2) electricity generation consumed only one-third of all energy resources (e.g., coal, oil, uranium). But even if nuclear electricity were to grow as proponents urge, tripling by 1990, it would make no material contribution to solving the oil crisis.

Electricity is peripheral to the oil crisis. It is one form of energy that is currently in substantial excess supply, and it can be produced from coal, our most abundant energy resource. Oil was not a major utility fuel even before the crisis, and after the "second shock" in 1979, when oil prices rose from under $15 to over $30 per barrel, utilities moved quickly to replace costly oil with much cheaper (and already existing) idle coal plants. Utility use of oil fell by nearly one-half between 1978 and 1981. By 1981 utilities were using only 7 percent of all oil consumed in the U.S., and the trend is still downward. No new oil-fired generating plants will be built, and most existing ones will be converted to coal or replaced within ten years, regardless of whether more nuclear plants are built. By the end of the decade electricity generation will account for only a few percent of total U.S. oil consumption.

There is only one circumstance in which nuclear power could help in solving U.S. oil problems. A shortage of nuclear power could conceivably cause an electricity shortage, which in turn would force consumers to rely more heavily on oil in, for example, home heating and manufacturing. In such a shortage situation ample supplies of nuclear-powered electricity would indirectly reduce the consumption of oil. But rather than running into shortages, utilities have been accumulating substantial surpluses of generating capacity, as construction programs have outpaced unexpectedly slow growth in demand. For the United States as a whole, generating capacity in 1980 exceeded maximum demand (on the hottest summer day) by 35 percent, whereas a 15–20 percent margin is considered adequate to assure reliable supply. Given the slowdown in electrical growth, present surpluses are unlikely to disappear much before 1985 even if cancellations and deferrals of nuclear plants continue at ecent high rates—rates which reflect utility efforts to adjust to the slowdown. In the onger run additional electricity, whether to substitute for oil or for other purposes, ould as well be provided by coal as by nuclear generation.

ʔalsely Equating Nuclear Power with Oil Savings

Despite all evidence to the contrary, nuclear and utility spokesmen continue to ·mphasize the oil "cost" of safety-related shutdowns of nuclear plants and licensing lelays instituted by the Nuclear Regulatory Commission. The falsity of this line of ·rgument, which implies a necessary connection between nuclear generation and oil ·equirements, was conclusively demonstrated by the experience of 1979: in the eight nonths following the accident at Three Mile Island, nuclear generation was 16 per- ·ent below the year-earlier level, an amount that would have raised oil consumption ·y electric utilities by 13 percent if the nuclear-oil equality were correct. Instead, oil onsumption actually *declined* by 17 percent, as utilities shifted generation toward ·oal (and to a smaller extent toward gas) to minimize the effects of sharply rising oil ·rices. This shift far more than offset any increases in oil consumption attributable o the nuclear shutdowns.

·he Importance of Coal

·hose who argue that nuclear power is necessary ignore the dominance of coal in ·lectric generation. Because almost five times as much electricity is produced by coal ·s by either oil or nuclear power, relatively small changes in coal generation can ·ffset large percentage changes in these other power sources. For example, between ·978 and 1980 oil generation declined by 33 percent and nuclear generation by about · percent; to replace both oil and nuclear required only a 15 percent increase in coal- ·red generation.

As the experience of the last few years shows, utilities will substitute coal for oil ·, for whatever reason, nuclear power is not a viable alternative. The reason is obvious: ·ven though most oil is consumed in areas where coal is relatively expensive, the ·rice of oil has risen to the point where oil costs two to three times as much as coal ·er unit of electricity generated, and the savings from switching from oil to coal are ·rge: over $100 million per year per 1000 megawats of capacity. Utilities are moving ·igorously to achieve these savings.

Even if nuclear construction programs were to be further delayed or halted alto- ·ether, it would cause only a temporary delay in the ongoing transition of electric ·tilities away from oil. With or without nuclear power, consumption of oil by utilities · one part of the oil problem for which a solution seems guaranteed.

The Air-Pollution Threat

Existing coal-fired generating plants are major sources of pollution, producing mor
than half of all man-made emissions of sulfur and nitrogen oxides (suspected to b
the primary cause of acid rain) and one-third of all smoke (''particulates''). To many
the threat of pollution from expanded coal generation is as unappealing as the danger
of nuclear power.

What needs to be recognized is that *no increase in pollution need occur, even i
all existing nuclear, oil, and gas generating plants were to be replaced eventually b
coal*—implying a 72 percent increase in coal-fired generation. This is because th
large pollution burden from coal plants is an artifact of past policies which paid littl
attention to air quality. Most older plants are many times dirtier than need be. Pol
lution could be reduced markedly by installing modern pollution-control equipmen
if all existing oil, gas, and nuclear plants were replaced by coal, and if all plan
(including existing coal plants) were required to meet the 1979 pollution standard
for new plants set by the Environmental Protection Agency, total emissions of oxide
would decline to about one-half of current levels and emissions of smoke to one
tenth.[46]

Lowering the pollution burden while shifting toward greater use of coal is ecc
nomically as well as technically feasible. At 1980 prices, conversion from oil to coa
would save enough in fuel costs in two and one-half years to pay for a technicall
advanced pollution-control system, including installation of a sulfur dioxide ''scrub
ber.''[47]

The Carbon Dioxide Threat

All fossil fuels release carbon dioxide (CO_2) when burned, and the rate at whic
industrial civilization is now burning these fuels is causing the level of CO_2 in th
atmosphere to rise at about 3 percent per decade. As the level continues to rise, mo
of the sun's heat will be trapped in the atmosphere, and world temperatures will ris
The extent of the likely rise in temperature and its effect on climate are subjects o
controversy among scientists, but there is wide agreement that the effects might b
very serious—possibly causing the polar ice caps to melt within the next 100 years

Although the absence of carbon dioxide is a point in favor of nuclear power, it
a minor point. Whether or not nuclear power is used will have only a modest effe
on the rate at which the world approaches its possible rendezvous with climatic disa
ter: the 500 large nuclear plants that might, under optimistic assumptions, be ope
ating in the world by the year 2000 would reduce the projected *rate* of CO_2 productic
by less than 10 percent. The atmospheric *level* of CO_2 would continue to rise almo
as fast as in the absence of nuclear power.[48]

Exaggerating Future Energy Requirements

Nuclear proponents often create an apparent need for nuclear power by assuming unrealistically high future requirements for energy. If electricity consumption in the United States were to double every ten years, as it did before the 1973 oil embargo, we would need four times the present amount of electricity by the end of the century. Meeting such a requirement solely with coal-fired plants would imply a sixfold expansion of coal mining and an eightfold expansion of coal-fired generation. But meeting the assumed expansion with nuclear would require building 1000 large (1000-megawatt) plants by the end of the century—a prospect no more attractive or feasible than an equivalent increase in coal-fired plants.

The obvious fact is that electricity use is not going to grow again at preembargo rates. Growth averaged only 3 percent per year from 1973 to 1980, less than half prior rates. Rising electricity prices, slowing population and economic growth, and spreading electronic innovation portend still slower growth in the future.

Even moderate expansion of the electricity sector is not inevitable. Future electricity requirements are not fixed by God or the utility industry. What we consume is open to public choice. If the risks or environmental consequences of expansion are considered unacceptable by the public, they can be avoided by increased emphasis on conservation—without any significant sacrifice of amenities. The potential for conservation through increased energy efficiency is very large, as has been documented in studies sponsored by the Harvard Business School,[49] industry,[50] the U.S. government,[51] and others.[52] Common sense and observation also support the conclusion that much electricity is being wasted. Examples of such waste are excessive lighting, overuse of airconditioning to substitute for inadequate building design and construction, and inefficient appliances and air conditioners. Assertions that electricity supply *must* grow assume implicitly that this waste must continue.

The Costs of Forgoing Nuclear Power

In the debate over whether the benefits of nuclear power justify the dangers, nuclear proponents often put their case in the most favorable light by comparing continued use of nuclear power to the most extreme alternative: *immediate* shutdown of *all* nuclear plants. There is no question that this would be expensive. Aside from possible short-term disruptions in electricity supply (which would be far less severe, at present, than nuclear proponents sometimes suggest),[53] about 60,000 megawatts of existing nuclear generating capacity would need eventually to be replaced, and billions invested in plants under construction would need to be written off.

Given its high cost, a total nuclear shutdown would only be ordered in the face of clear evidence (most probably provided by a nuclear catastrophe) that continued oper-

ation of nuclear plants posed a severe threat to public safety. In this context neithe
the oil nor dollar costs presently appear unmanageable—although this will change a
dependence on nuclear power rises.

As of 1980 there was sufficient nonnuclear generating capacity available natior
ally to meet the maximum requirement (peak demand) for electricity.[54] The cost c
replacing one-half the 1980 nuclear generation with oil and one half with coal
would have been about $7 billion (8 percent of the national electric bill and 2 percen
of the national energy bill). Oil consumption would have increased by about 600,00
barrels of oil per day (3.5 percent of total consumption and 9 percent of imports).

Of more interest than the extreme case of a nuclear shutdown is the cost of fo
going nuclear power in the longer run. What would be the cost to the nation if n
more nuclear plants beyond those already under construction were to be built? Althoug
all answers must be qualified because of the great uncertainties pervading the energ
field, electricity from nuclear plants now appears likely to be more expensive tha
that from coal-fired plants.[56] Thus if the nation builds no more nuclear plants, it
likely to save money.

Nuclear proponents of course argue that nuclear energy will be cheaper than coa
but even if they were correct, the maximum possible savings from nuclear would ne
be sufficient to materially affect the overall costs of energy to the country. The debate
over whether or not nuclear power will be more or less expensive than coal pow
revolve around differences in cost of at most one cent per kilowatt-hour. Suppose th
country were to triple the size of the nuclear sector in the next ten years, by addin
100 large new plants, and that the new plants were to save one cent per kilowatt-how
compared to new coal plants. This large nuclear program would only reduce the co
of the national energy bill by the same amount as would a *two-cent*-per-gallon redu
tion in the price of petroleum products.[57]

The energy crisis is dominated by oil, and nuclear power cannot make muc
difference. Nuclear power appears important to electric utilities and nuclear man
facturers because they have billions of dollars invested in it. But when viewed in th
overall context of the energy crisis, nuclear power is, in the vernacular of my part c
the country, mighty small potatoes.

The Effects of a Nuclear Phase-out

ALAN S. MANNE AND RICHARD G. RICHELS

Alan S. Manne is a professor of operations research at Stanford University. Richard C
Richels is a technical manager for the Electric Power Research Institute, an independe
research organization for the utility industry. Before moving to EPRI in 1976 Dr. Rich
was a consultant to the Rand Corporation and the National Science Foundation. Drs. Man

The individual authors are solely responsible for the views expressed in this essay.

nd Richels's energy projections have been used for a number of energy studies, including he Committee on Nuclear and Alternative Energy Systems (CONAES) report. For this essay ie authors use a computer-based energy-economic model to predict the effects of a nuclear ioratorium on the U.S. economy. Contrary to what Dr. Taylor has suggested, Drs. Manne nd Richels believe that abandonment of the nuclear option over the next fifty years could e quite costly, with a possible trillion-dollar loss to the American economy.

ince well before the accident at Three Mile Island, civilian nuclear energy policy has een virtually stalemated by costly delays, regulatory changes, and general inflation. 'or the immediate future this stalemate may continue. Nuclear energy is a highly ivisive issue, and it is not easy for democracies to arrive at a political consensus in iis area. Eventually, however, this stalemate is likely to end.

Concerns about weapons proliferation, reactor safety, and radioactive waste dis-osal will undoubtedly continue to influence national decision-making. Therefore we o not mean to suggest that the debate can be reduced to simple economic compari-ons of various energy alternatives. But we do feel that society's willingness to accept r reject risks should depend, at least in part, on the economic consequences. These ecisions have a direct bearing on our standard of living and the health of the country.

In this essay we will compare the cost of electricity and its impact on the nation's conomy under two rather different scenarios, one with and the other without nuclear ower. Our purpose is to provide an overall analysis of the domestic U.S. cost of ejecting the nuclear option. The costs would probably be higher if other nations were imultaneously to follow this course.

lternatives to Nuclear Power

Io technology assessment can be conducted in isolation. For example, the economic ttractiveness of nuclear power will depend as much on the detrimental environmen-l effects of burning coal as it does on the construction costs of nuclear power plants. the principal alternatives to nuclear power were cheap, plentiful, reliable, and lean, the choice would be easy. But there are no easy choices. Each alternative has s own difficulties, and these must be considered in any realistic assessment of future nergy policies.

Within this century as we gradually exhaust our oil resources, the principal short-rm alternatives are coal, nuclear, and conservation. Coal, a well-known and reliable orm of energy, could probably provide the bulk of our electricity for the remainder f this century. The costs of coal-fired electricity, even in Europe and Japan, proba-ly do not exceed those of nuclear power by more than 25 percent. Phasing out uclear power would be economically tolerable if there were no practical environ-iental constraints to the expansion of coal use.

Coal, however, is no panacea. The environmental effects of routine emissions

from coal plants are highly uncertain. Even with scrubbers, which can extract ove
90 percent of the sulfur emissions from coal plants, studies show that anywhere from
zero to eight fatalities a year can be attributed to the sulfur emissions produced from
just one coal-fired plant.[58] Estimates of various lung diseases due to air pollution ar
even less definitive, with high and low estimates differing by two orders of magni
tude.[59] These large uncertainties are due to our lack of knowledge of complex bio
logical and atmospheric effects.

In the long term the main threat from coal burning may be to the earth's climate
The burning of fossil fuels produces carbon dioxide (CO_2), which acts like a green
house to trap the sun's rays. Some investigators suggest that a substantial increase i
the concentration of CO_2 in the atmosphere could increase the average surface tem
perature of the earth by as much as two to three degrees over the next 100 years.[6]
Although there are enormous uncertainties in this calculation, the fact remains tha
such a hypothetical rise in the earth's temperature *could* have potentially catastrophi
effects, causing flooding and other natural disasters.

The second short-term alternative to nuclear power is conservation. Given th
economic incentives and enough time, conservation can provide significant reduc
tions in energy demand. Some critics of nuclear power have gone so far as to sugges
that increased conservation efforts alone can eliminate our need for nuclear powe
This argument, however, is misleading. If enough energy could be replaced easil
by capital and labor (e.g., through insulation to reduce residential heating require
ments), then conservation could offset the effects of phasing out nuclear energy. Bu
the opportunities for low-cost conservation are limited, and will in any event b
needed to offset future declines in the production of conventional oil and gas. Add
tional conservation will be far more costly. These costs are difficult to determine. I
the jargon of economists, this is a disagreement over the numerical value of th
"elasticity of substitution" between energy, capital, and labor. Seemingly sma
changes in this elasticity can produce major changes in the economic effects of phas
ing out nuclear power.[61]

What about other alternatives, such as solar technology? Over the long term sola
and even fusion power* offer the hope of a nearly unlimited supply of electricity
Although the critics of nuclear often claim that such advanced sources of electrici
are just around the corner, it remains uncertain when these sources will in fact b
available. Serious technical and economic problems must be overcome before *any* c
these advanced systems can become a major energy source.

In discussing the prospects for solar electricity, a study conducted by Resourc
for the Future commented that "large-scale direct solar sources of electricity see
unpromising for the United States within the next couple of decades, even assumir
improvements in costs of thermal collectors of photovoltaic cells."[62] When consi
ering the role for fusion in our energy future, this same report was even less optimi

---

* Massive quantities of energy can be derived either by splitting the uranium atom (fission) or by fusi
hydrogen nuclei together (fusion). See Chapter 6.

ic: "Since even scientific feasibility has not yet been demonstrated for fusion power, it is somewhat premature to discuss detailed reactor economics."[63]

It is not our purpose to present an exhaustive technical analysis of the long-run alternatives to nuclear power. Nor would it be practical. There is little doubt of the nation's ultimate ability to develop alternative energy sources, but the issue is timing. The economic consequences of a "nuclear phase-out" decision are critically dependent on how long we might have to wait for viable substitutes. Technical breakthroughs are always possible, but it would be a mistake to plan our energy future on the assumption that new sources of electricity will soon make a significant contribution.

Two Energy Futures

In an attempt to determine the costs of rejecting nuclear power in the United States, we will explore two very different energy paths: one with nuclear power and one without. For the first, we assume that nuclear power expands at 10 percent per year, reaching 250 gigawatts of power at the turn of the century, and then expands at 5 percent per year thereafter. This would represent much slower growth rates than were projected in the early 1970s, but is consistent with the recent slowdown in GNP growth and energy use.

By contrast the nuclear phase-out scenario assumes that no new plants go on line after 1980. To avoid economic disruptions caused by this large shift, we assumed that the existing reactors are phased out gradually over the balance of their thirty-year lifetime.

Realizing the complexity of a nuclear phase-out, we have used a computer model to calculate its economic impact. Of course no computer can deal with all the uncertainties of a fifty-year projection. Nonetheless we feel that this provides a more systematic approach and is open to more outside scrutiny than the guesswork and intuitive judgments that often characterize the nuclear controversy. This allows us to make systematic variations in factors (such as the rate of coal consumption) which have a major impact on the overall costs of a nuclear phase-out.

Our computer model is called ETA-MACRO.[64] It was designed for energy technology assessment, and has served as a basic tool in several national cost-benefit analyses.[65] Because of the major uncertainties in the nuclear controversy, it would be naïve to give much weight to any single projection from such a model. But by varying the assumptions one at a time we can see the economic impact of different scenarios and develop some understanding of what it would mean if there were a systematic long-range policy to phase out nuclear power.

In this version of ETA-MACRO, baseload electricity is produced by coal-burning power plants, nuclear fission, and an advanced electric technology.[66] "Nuclear fission" includes both the currently available light-water reactor (LWR), and the more advanced fast breeder reactor (FBR) after 2015. The "advanced electric technology" could be either solar or fusion. It is assumed that these are available by 2000, but a twice the energy costs of conventional units.

Fig. 5–2 compares our cost assumptions for the four main types of electric powe plants.[67] Notice that the price of coal-fired electricity is much more sensitive tha that of nuclear power to the cost of fuels. At a coal plant the fuel bill accounts for a third or more of the costs of the electricity generated. Natural uranium, by contrast constitutes less than 10 percent of the costs in an LWR.

Breeders are even less dependent on fuel costs. But a breeder reactor produce more fissile material (plutonium) than it consumes, and so it is able to extract mor than 60 percent of the thermal energy available in uranium—as compared to the 1 o 2 percent extracted by the LWR. Hence breeder development would offer consider able fuel savings over LWRs.

Fig. 5–2 provides some immediate insights into the relative economic attractive ness of these four electric technologies. When uranium costs $50 per pound and coa costs $1.50 per million BTUs, the LWR has only a slight cost advantage over coal fired units in the U.S. The LWR has a significant advantage only when environmenta or other constraints raise the price of coal utilization.

According to our estimates the LWR (without reprocessing) is more economica than the breeder reactor, provided that uranium prices remain below $120 per pound It is only when low-priced uranium resources become depleted that the breeder reac tor is significantly more attractive.

The calculations that follow are based on an optimistic outlook for the supplies o *non*electric energy derived from oil, natural gas, and unconventional sources. It i supposed that limited quantities of synthetic fuels may be derived from coal and shal oil, but that virtually unlimited supplies may come from other sources (heavy oils solar heating, biomass conversion, etc.). We assumed that international oil price will double between 1980 and 2010, and that these unconventional sources will there after be competitive with imported oil. In effect this assumption rules out the possi bility that nuclear energy would reduce our vulnerability to the threat of sudden disrup tions in the price or availability of foreign oil. That issue is not explored here.

In ETA-MACRO, electric and nonelectric energy are used in combination wit capital and labor to produce the economy's GNP. (We assume that at constant energ prices, the GNP could grow at an annual rate of 3.2 percent between 1980 and 2000 Electric energy, nonelectric energy, and conservation are combined so to maximiz overall economic benefits. When one or more sources of energy are limited, there i a reduction in economic productivity.

Costs of a Nuclear Phase-Out

For this application of ETA-MACRO, four alternative scenarios have been defined which differ in terms of *both* coal and nuclear energy. Under scenarios I and II it i

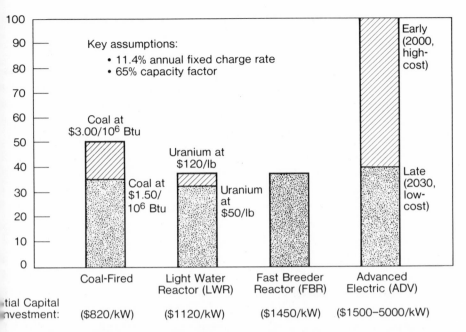

Fig. 5–2   Electricity Cost Comparisons. The graph compares the cost of electricity from various sources at different fuel prices. When coal costs $1.50 per million BTUs (approximately its present price), then coal-generated electricity will cost 35 mills per kilowatt-hour. If coal doubles in price, then coal-generated electricity rises to only 50 mills per kilowatt-hour. Thus doubling the price of coal does not double the price of coal-generated electricity. Electricity from light-water reactors is even less dependent on the cost of fuel (uranium) than is coal-generated power. The cost of electricity from a breeder reactor is virtually independent of the price of fuel. As the graph illustrates, LWRs have a slight advantage over coal only if uranium costs $50 per pound and coal costs $1.50 per million BTUs. The breeder only becomes attractive if the price of uranium exceeds $140 per pound. Advanced sources of electricity (solar and fusion) should be very expensive when they become available in the year 2000 (100 mills per kilowatt-hour). Their price should not drop to 40 mills per kilowatt-hour until the year 2030.

ssumed that coal use will be limited by external factors. By contrast, no coal limits re imposed in scenarios III and IV. In these cases it will be seen that there is an normous potential demand for coal. Scenarios I and III are "with nuclear," whereas and IV refer to a "nuclear phase-out."

Our coal limits are meant to reflect realistic constraints upon the annual rate of oal consumption for all purposes taken together: coal-fired electricity, direct uses or coal, and for synthetic fuels. This upper bound would allow coal consumption to ouble between 1980 and 2000. (Between 1970 and 1980 coal consumption expanded y only 17 percent.) There is some question as to whether coal *production* can be xpanded rapidly enough to keep up with domestic and export demands. Eastern coal roduction may be limited by manpower and by union activities. Western coal faces transport and infrastructural problems. These production constraints may prove to be an even more serious immediate obstacle than the health and environmental issues.

Electricity Prices, Coal Consumption, and Energy Demands   Fig. 5–3 compares the future electricity prices that are needed to bring supplies and demands into balance. First, compare the price paths when coal is constrained (Fig. 5–3A). Both coal and electricity prices rise whenever coal demands bump up against a utilization constraint. The price rise is significantly higher in the nuclear phase-out scenario. In this case coal would be diverted from synthetic fuels production to replace some of the electricity that would otherwise have been produced by nuclear units. This puts heavy pressure on coal and hence on energy prices in general.

Note that the significance of the coal constraint diminishes over time, however. New electric and nonelectric supply options and the shift away from energy-intensive techniques and lifestyles eventually combine to cause a post-2000 decline in electricity prices.

The strategic role of coal becomes apparent from scenarios III and IV (Fig. 5–3B). Here it is assumed that unlimited quantities of coal can be delivered at prices of $3 per million BTUs (versus $1.50 in 1980). Under this assumption a nuclear phase-out leads to relatively small impacts on electricity prices. The ceiling on the delivered price of coal places a 50 mills per kilowatt-hour lid on the price of electricity.

Table 5–1 compares coal consumption and energy demands for each of our four scenarios. If we assume that virtually unlimited resources of coal are available, a nuclear phase-out would lead to a *tripling* of coal demand between 1980 and 2000, and even larger subsequent increases (scenario IV). However, when coal is constrained, a nuclear phase-out also takes its toll through the cost of reducing demands in response to higher energy prices.

Macroeconomic Costs   Let us now examine the economy-wide consequences of phasing out nuclear power. Higher energy prices lead to a productivity slowdown, and conservation becomes increasingly more expensive. The economic impact may be measured in terms of the GNP loss, but it is more accurate to measure this loss in terms of the aggregate consumption of goods and services. Fig. 5–4 describes the economy-wide effects when coal utilization is constrained. Although the losses are small in percentage terms, they are nonetheless significant on an absolute scale during the post-2000 period. In the year 2030 alone consumption losses would be over $600 billion (in 1979 dollars). When these costs are added over all years from 1980 to

Fig. 5–3 Electricity Price Comparisons. A. In scenarios I and II we assumed a ceiling on the future price of coal. Notice that the price of electricity is significantly lower with nuclear power than without. **B.** In scenarios III and IV we assumed *no* ceiling on the price of coal. Notice that a nuclear phase-out when coal consumption is unlimited has less of an effect on the price of electricity than a nuclear phase-out with restricted coal consumption.

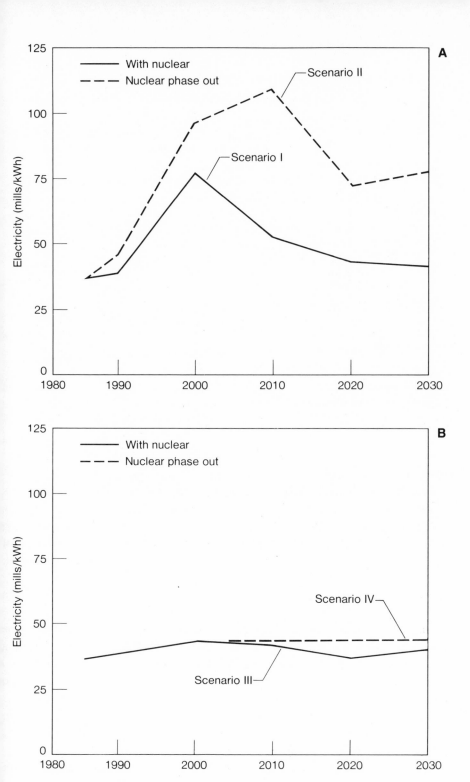

TABLE 5–1   COAL CONSUMPTION AND ENERGY DEMANDS[a]

Scenario	Coal Consumption (quads)[b]		Electricity Demand (trillion kWh)		Total Energy Demand (quads)[b]	
	2000	2020	2000	2020	2000	2020
Scenario I (with nuclear, limited coal)	34.0[c]	50.0[c]	4.4	8.1	110.8	161.3
Scenario II (nuclear phase-out, limited coal)	34.0[c]	50.0[c]	3.5	5.3	102.0	131.7
Scenario III (with nuclear, unlimited coal)	42.2	62.7	5.2	9.4	117.0	171.9
Scenario IV (nuclear phase-out, unlimited coal)	54.5	95.7	5.2	8.6	116.2	161.1

[a] Actual values for 1980: coal consumption, 16 quads; electricity demand, 2.4 trillion kWh; total energy demand, 76 quads.
[b] A quad is a quadrillion British Thermal Units (BTUs).
[c] Upper bound on coal consumption.

2030, and are discounted to 1980 at a 5 percent annual rate, the total economic loss associated with a nuclear phase-out strategy is nearly $1200 billion.[68]

In the unconstrained coal scenarios there would be far lower macroeconomic losses from phasing out nuclear power. On an absolute scale, however, these would still be significant—nearly $300 billion at a 5 percent discount rate between 1980 and 2030.

Summary and Conclusions

During the next several decades conservation and coal represent the principal alternatives to nuclear power. Neither of these alternatives is cheap. Each will become increasingly expensive if it is the sole option available. In any event future energy supplies are likely to remain tight, and energy consumption will have to grow more slowly than the GNP. A nuclear phase-out would further slow down the growth or productivity of the U.S. economy. Depending on what is assumed with respect to coal utilization, the macroeconomic cost of such a phase-out could vary between $300 and $1200 billion (discounted at 5 percent per year between 1980 and 2030).

The $1200-billion estimate is applicable if the use of coal is constrained by health environmental, or global climatic consequences. The cost might, however, be as low

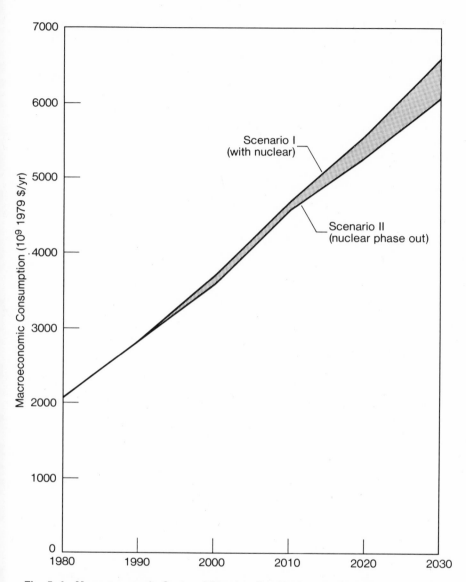

Fig. 5–4 Macroeconomic Costs of Phasing Out Nuclear Power When Coal Use Is Constrained. The additional cost to the economy from a nuclear phase-out (shaded area) would reach $600 billion in the year 2030 (assuming restrictions on the availability of coal). When these additional costs are added up over a fifty-year period, the total cost would exceed $1200 billion.

as $300 billion if coal use can be expanded greatly. Until the basic scientific uncertainties on coal combustion are reduced it will be very difficult to narrow this range further. What is certain, however, is that there are significant economic costs that should be factored into society's decisions to accept or reject the risks of nuclear energy.

6

Beyond Light-Water Reactors

Introduction

We are slowly running out of uranium. Even if light-water reactors can weather the present controversy, commercial nuclear power as we know it has a limited life because this country will have exhausted the bulk of our known uranium reserves within perhaps forty years.*

The debate over which energy technologies to pursue is not an academic one, because we are quickly depleting conventional fuel reserves. By the year 2020 oil, which fueled the expansion of the industrial nations of the world for an entire century, will be scarce and extremely expensive. Low-cost uranium resources, enough to fuel at present prices perhaps 700 light-water reactors, could be nearing depletion. Coal, and to some extent natural gas, will continue to be key fuel sources during this period. But although the United States has some of the largest coal supplies in the world, enough to last several hundred years at present mining rates, strict environmental controls and limited transport infrastructure will limit the expansion of coal mining.

Nuclear power, therefore, stands at a crossroad. Decisions must be made *now* if alternative energy technologies are to be tested and brought on line by the year 2020. Yet while many scientists, business executives, and politicians are convinced that the future lies with the "second generation" reactors such as the breeder or fusion reactor, which have not yet been commercially developed, a growing number of people are pursuing nonnuclear alternatives: wind, solar, and hydropower. In theory it is possible to heat our homes or operate our cars using energy ranging from synthetic fuels to walnut shells. But we need to ask ourselves which options are the cheapest, the most desirable, and the best able to provide substantial amounts of energy in the shortest period of time.

At hundreds of "energy fairs" across the country the benefits of renewable non-nuclear resources have been proclaimed by students, farmers, environmentalists, and amateur scientists who display their backyard windmills, solar collectors, composting

---

*Large uncertainties exist in calculating the total supply of uranium ore, the rate and efficiency with which uranium will be consumed by light-water reactors, the number of reactors built, the future rate of growth in electricity consumption, and other crucial factors. Consequently some energy analysts predict uranium might remain competitive as a fuel source even after 2050 *if* current trends continue to slow down electricity growth and limit nuclear expansion.

toilets, and gasohol cars. Through skits, workshops, lectures, and colorful, eclectic displays, an image is presented of a pollution-free America with decentralized energy, which could replace the threat of meltdowns and radiation contamination. Accompanying the placards and posters is a clear philosophical message: nuclear power is the child of the same energy monopolies that gave us pollution in the first place. With solar energy and other renewable resources we can once again take control of our lives and free ourselves from centralized utilities and energy monopolies. "People before profits" reads the literature.

Conferences sponsored by the Atomic Industrial Forum and the American Nuclear Society emphasize that these energy fairs offer a solution that's about as effective as "a mosquito bite on an elephant's fanny," in the words of the editor of *World Oil*. Solar heaters may be fine for summer vacation homes, but are hardly a reliable source of energy for the largest energy-consuming industrialized nation in the world. Industrialists declare that nuclear power deployed in breeder and fusion reactors can be an extension of the same technology that gave the United States one of the highest standards of living in the world. They maintain that the insatiable demands for energy in this country can be met only by a larger commitment to nuclear power.

Between these two groups stands a wide range of pragmatists with no ideological commitment to any particular energy source. They are people—scientists, businessmen, economists—who have evaluated the pros and cons of many energy sources and now support one or several alternatives to light-water reactors. This chapter presents four of these alternatives. The first two options are nonnuclear, involving a variety of renewable energy sources, including solar energy and improvements in energy efficiency. The second two options are nuclear, involving the development of the breeder and fusion reactors as a means of extending the lifetime of nuclear energy. Incredibly, each of these alternatives promises an inexhaustible source of energy if commercially developed.

Soft Energy Paths

In the past few years, the United States has gotten more than fifty times as much new energy from more efficient use as from all expansions of energy supply combined, and of those expansions, probably the largest has come from renewable sources.

—Amory Lovins, policy advisor,
Friends of the Earth

Amory Lovins is the physicist who coined the phrase "soft energy path" and introduced the concept of using technologies that are appropriate for each energy task. In Lovins's lexicon, "hard energy" represents the mammoth, centralized, often electrified energy technologies—including coal, oil, and especially nuclear—that waste and eventually exhaust limited resources. According to Lovins and his co-author Hunter Lovins, nuclear power is a white elephant, symbolizing an outdated philosophy of supplying ever more energy in higher quality forms from fewer, bigger sources, regardless of whether that is the cheapest way to provide each energy service.

In "soft energy paths," on the other hand, we would use energy far more efficiently, and switch gradually to diverse renewable sources such as solar heat, liquids from biomass wastes, hydro, and wind. In combination these sources, the Lovinses argue, could save money, rapidly eliminate both oil imports and nuclear power, and *reduce* total energy use despite (indeed, as a main way of achieving) a *rise* in national wealth. This prospect requires that we take advantage of new technologies to raise energy efficiency. For example, Amory and Hunter Lovins contend that just fixing up the motors in America's factories to cost-effective levels of efficiency would save enough electricity to replace every nuclear reactor in the country.

Proponents of energy efficiency have been accused of trying to move our economic system toward socialism; their contention that efficient use of energy can make nuclear power obsolete has been seen as an effort to slow growth across the board. The Lovinses' contribution is thus all the more provocative because it is based on private enterprise, free-market economics, and an assumed goal of rapid economic and industrial growth worldwide.

Where will Americans get the electricity that is needed if not from the hard technologies? It is for readers to decide whether the Lovinses have answered this question satisfactorily.

Solar Energy

Many people still assume that solar energy is something for the future, awaiting a technological breakthrough. That assumption represents a great misunderstanding, for active and passive solar heating is a here-and-now alternative to imported oil.

> —Modesto A. Maidique, associate professor of engineering management, Stanford University

With these words, Professor Maidique summarizes his analysis of solar energy alternatives. He believes that solar is a technology that is commercially feasible and that perhaps 20 percent of all energy could be derived from solar (broadly defined) by the year 2000 if necessary.

Both the solar collector and solar photovoltaic cells are alternatives today to imported oil. Tens of thousands of homeowners have installed solar collectors, which are often simply pools of water sealed under glass plates designed to collect the rays of the sun. Thousands of others have built passive solar (energy efficient) homes. Though a solar collector cannot produce electricity, it can adequately heat hot water and space-heat homes and thus save on increasingly expensive conventional energy sources. A more sophisticated utilization of solar energy, which is still too expensive for most residential or commercial use, is derived from panels of solar (photovoltaic) cells which capture the sun's energy and convert it directly into electricity. Development of this technology originated with the space program where expensive solar panels are used to power satellites.

The future promises forms of solar power that are even more exotic. Biomass (the energy derived from burned waste plant material) may one day power entire cities. Solar satellites (which convert the sun's rays into microwaves) could one day transmit abundant energy from outer space to earth.

Solar energy in all its forms, though presently supplying 6 percent of our total energy needs, still faces institutional, economic, technical, even ideological barriers. One problem is that many forms of solar energy are not yet marketed as a complete system to the consumer. Effective distribution, promotion, and service systems are just beginning to emerge. Some experts think that if solar energy is to become widespread, then big business must step in to mass produce and market most solar technologies.

Professor Maidique takes a hard, businessman's look at solar energy's potential and comes up with some surprising observations. Perhaps solar technology is not as farfetched as its detractors claim, nor the panacea that its supporters would like.

Fusion

Designing a nuclear fusion reactor . . . is a little like planning to reach heaven; theories abound on how to do it, and many people are trying, but no one alive has ever succeeded.
—David J. Rose, professor of nuclear engineering, MIT

Fusion energy is what powers the sun, the stars, and the hydrogen bomb. Whereas fission splits apart the uranium nucleus to release energy, fusion is the joining together of nuclei of hydrogen atoms in order to release energy. The fuel for fusion reactors can be derived from ordinary seawater, available in virtually unlimited quantities, and the reaction produces significantly less waste than does fission.

Does this sound too good to be true? Although nuclear engineers like Dr. Rose point out that commercial fusion is decades away, fusion advocates maintain that a demonstration fusion reactor is finally within our grasp. Scientists in three countries (the United States, England, and Japan) are closing in on controlled fusion: each country is building a demonstration fusion reactor which it is hoped will break even— that is, produce as much energy as is consumed by the reaction. Although the attainment of commercial fusion is an engineering task of mindboggling complexity— rather like bringing a piece of the sun down to earth—a major experimental fusion device is being built at a laboratory at Princeton University, where scientists are confident that we can break even with today's technology. The remaining problems are of a technical (rather than scientific) nature, so that these plants, if successful and economical, might be built commercially and brought on line in the next few decades.

Some engineers still wonder whether these multi-million-dollar devices will ever generate electricity. But Dr. Stephen O. Dean, president of Fusion Power Associates, a group of engineering and industrial firms engaged in fusion development, explains why fusion energy might very well keep our lights burning in the next century.

Breeder Reactors

I like the sun [said physicist Hans Bethe]. After all, I found out how the sun works.* But after looking at the available evidence I do not believe that the sun will solve our problems—certainly not in the twentieth century.

Dr. Bethe warns that though solar power is appealing, the United States should not count on it as a near- or middle-term future energy option. Instead Dr. Bethe and his co-author, Dr. Robert Avery, prefer the breeder reactor, pointing out that the breeder is a *proven* technology, while solar power for generating electricity requires major scientific and economic developments.

Like the fabled philosopher's stone that turned lead into gold, the breeder reactor turns waste uranium-238 into plutonium, which could vastly extend the lifetime of nuclear power. The importance of this seemingly magical energy source was recognized back in the 1940s when one of the great nuclear pioneers, Enrico Fermi, claimed that "the country which first develops a breeder reactor will have a great competitive advantage in atomic energy."

That country may not be the United States. Although most administrations since Truman have given their blessing to the breeder reactor (President Nixon claimed in 1971 that "Our best hope for meeting the nation's growing demand for clean energy lies with the fast breeder reactor"), the Carter administration deferred the Clinch River Breeder Reactor Demonstration Project and slowed breeder development. Although President Reagan is proceeding with the Clinch River reactor, funding for the project barely passed through Congress in 1981 and the reactor is not expected to go on line until the late 1980s. Meanwhile critics of nuclear power are vehemently opposed to the breeder for two reasons. First, if the industry proceeds with its plans, in this "plutonium economy" there would be perhaps several hundred breeder reactors in the U.S., each requiring the reprocessing of spent fuel and transport of large plutonium shipments, which could be hijacked and diverted for weapons use. Second, unlike the process in a light-water reactor, there is concern that, because the core contains plutonium, a loss of sodium coolant could trigger a small nuclear explosion—the "criticality" accident—the force of which could be comparable to an explosion of perhaps (but no more than) several hundred pounds of TNT, arguably enough to release plutonium and other radioactive materials into the atmosphere.

Dr. Bethe and Dr. Avery argue that a modern industrial society can successfully

* Dr. Bethe was awarded the Nobel Prize in physics in 1967 for discovering the specific nuclear reactions that produce energy in the sun and other stars.

control this technology, and explain that criticality accidents would be much less severe than originally suspected, due to additional safeguards and more realistic assumptions. Dr. Avery, director of the Reactor Analysis and Safety Division of Argonne National Laboratory, has been involved in studying reactor safety for a number of years, as has Dr. Bethe, who is a professor of nuclear physics at Cornell University.

Soft Energy Paths*

AMORY B. LOVINS AND L. HUNTER LOVINS

Amory B. Lovins is a consultant physicist, a former Oxford don, and in 1980–1981 a member of the Energy Research Advisory Board of the U.S. Department of Energy. His wife and colleague L. Hunter Lovins is a lawyer, sociologist, political scientist, and forester. They serve jointly as policy advisors to Friends of the Earth, Inc., and in 1982 were Luce Visiting Professors of Environmental Studies at Dartmouth College and taught at the University of Colorado at Boulder. The Lovinses are active in energy policy in over fifteen countries, have published eleven books, and advise a wide range of governments, international organizations, and academic and industrial clients.

The nuclear enterprise is dying of an incurable attack of market forces, not only in the United States but in every market economy in the world. Nuclear power, presently delivering to the United States about half as much energy as wood, is rightly perceived by investors in the marketplace to be an uneconomical and risky investment, regardless of the subsidies and guaranteed returns now offered. It is not surprising, then, that official forecasts of how much nuclear capacity we would have in the year 2000 fell during 1973–1979 by a factor of five for the world and eight for the United States. Nuclear power probably never *will* get out of the firewood league.

How do we know that the nuclear collapse results from market forces, rather than from protests, interventions, inept regulations, and the nuclear industry's other favorite scapegoats? Fig. 6–1 shows the U.S. government's range of forecasts of the U.S. nuclear capacity which was to have been installed by 2000 (and the capacity currently installed). As experience with the technology grew, the forecasts declined. Compare this with Fig. 6–2 which shows the corresponding official forecasts for a country (Canada) which does not allow interventions or licensing hearings, where the main safety documents are secret, and where the government simply builds reactors wherever it chooses. Finally, Fig. 6–3 traces the nuclear forecasts for a country with unregulated electric utilities that charge whatever prices they want: West Germany.

---

* See A. B. Lovins and L. H. Lovins, *Energy/War: Breaking the Nuclear Link* (San Francisco: Friends of the Earth, 1980; New York: Harper and Row Colophon, 1981); and (with F. Krause and W. Bach) *Least-Cost Energy: Solving the CO$_2$ Problem* (Andover, Mass.: Brick House, 1982), both of which provide extensive documentation for the arguments summarized in this essay.

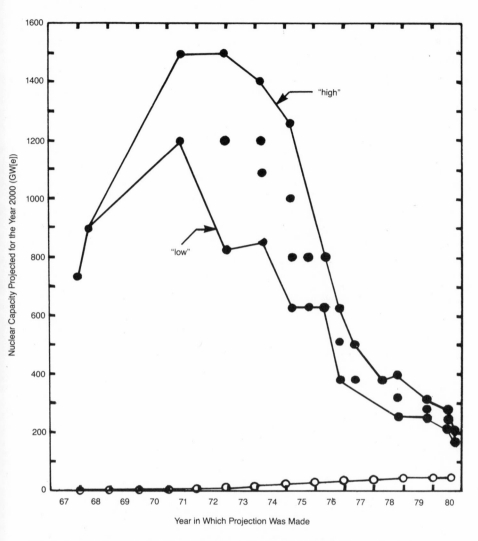

Fig. 6–1 U.S. Nuclear Power Projections for the Year 2000. The open circles represent the *actual* growth in U.S. nuclear generating capacity from 1967 to 1980 (in gigawatts). Note the narrowing gap between the line they form and the solid circles, representing the range of official *forecasts,* issued in those years, of what U.S. nuclear capacity would be in the year 2000 [forecasts by AEC/ERDA/DOE].

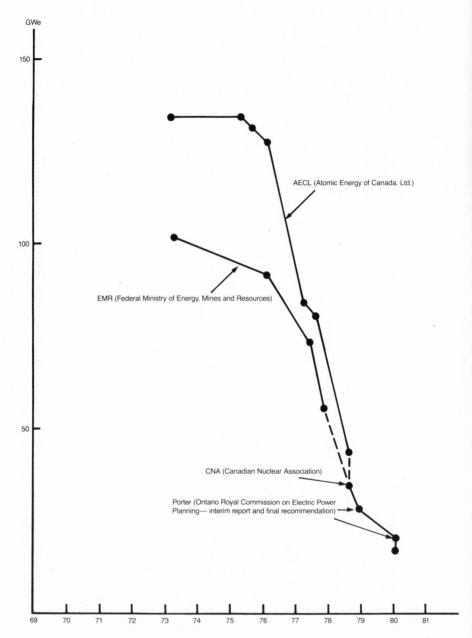

Fig. 6–2 Official Canadian Nuclear Power Projections for the Year 2000. Even without the burden of licensing hearings, environmental impact statements, other regulatory burdens or even public access to the basic documents, the decline of the forecast for Canadian nuclear capacity for the year 2000 is essentially identical to that of the United States.

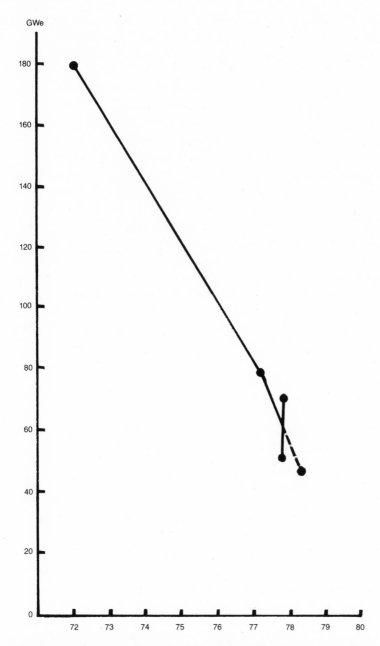

Fig. 6–3 Projected Nuclear Capacity of the Federal Republic of Germany in the Year 2000. Official forecasts of West German nuclear capacity in the year 2000 show a similar decline even though German utilities are unregulated. The estimate for 1978 by the U.S. government; no German forecasts were published after 1977.

The indistinguishability of the three patterns clearly shows that the collapse is a fundamental economic effect, not an American political idiosyncracy. Similar patterns of collapse prevail worldwide except in such centrally planned economies as the Soviet Union and France.

There is a common misconception, furthered by many nuclear advocates, that the only substitute for reactors is increased dependence on coal-fired power stations. Coal plants, which today provide Americans with nearly five times as much electricity as either nuclear or oil plants, are indeed likely to outpace nuclear plants at lower cost. But arguing that coal is cheaper and more available than nuclear energy is hardly an endorsement. Coal, even with its air pollution reduced as impressively as Vince Taylor describes, has serious social and environmental side effects, as the nuclear advocates are quick to point out. Fortunately, arguing coal versus nuclear is a bit like debating the relative economic merits of Iranian as against Russian caviar—to caviar fans, interesting, but not relevant to most of us. We have energy options *other* than more giant power stations, whether fueled by nuclear or coal. Comparing the costs of coal and nuclear as if we had no other way to meet energy needs ignores choices that are safer, cheaper, and cleaner than either. To understand this, however, we need first to ask some basic questions: How much energy do we actually need? What *kinds* of energy do we need? Where can we get it? And how can such an energy future be implemented?

How Much is Enough?

It is fashionable to suppose that we are running out of energy, and that the solution is to get more of it—as if the more energy we use, the better off we are. But asking where to get more energy begs the question of how much we really need. Most energy analyses assume that because until the 1980s we have tended to use an increasing amount of energy, future energy needs will be enormous. But merely projecting past trends ignores both economic and technical advances. How much energy we will need tomorrow depends not on how much we used yesterday, but on how much energy it will take to do what we want to do in the future.

The amount of energy it takes to make steel, drive a car, run a sewing machine, or keep you comfortable in your home depends on how cleverly we use energy. When oil was cheap and plentiful we could afford to use more energy and less skill. But as energy prices have risen it has become worthwhile to put more effort into using less fuel. It is now cheaper, for example, to double the efficiency of most industrial electric motors through better sizing, coupling, and controls than to buy electricity to run the old ones. This technological improvement, if systematically implemented, would save approximately 75 gigawatts of installed capacity in the U.S., more than enough to replace the entire U.S. nuclear power program. Similarly, cost-effective improvements in the lighting of buildings would provide the same standard of illumination for a third of the energy. Efficient household appliances, already designed but not yet marketed, would perform the same work using only a quarter as much energy as appliances use today.

Surprised? Let's consider a design example. Remember the old, mainly prewar refrigerators that had the motor mounted on top? Those motors were about 90 percent efficient. Today they are nearer 60 percent efficient and are mounted underneath, so the heat rises to where the food is. Your refrigerator probably spends half its effort taking away the heat of its own motor. When you open the door the cold air falls out and the metal frosts up inside, so most refrigerators have electric heaters inside that turn on occasionally to melt the frost, and more heaters to keep the door gaskets from sticking. The manufacturers have also kept trying to make the inside bigger but not the outside, so they skimped on insulation, and now the heat comes right in through the walls. The refrigerator is often installed next to a stove or dishwasher, which heats it up some more. It's hard to think of a dumber way to use electricity. Modern design can provide a refrigerator as elegant as the best now on the market—and one that can keep the same food just as cold, just as conveniently, using only a sixth to a twentieth as much electricity as now.

Our whole economy is like that wasteful refrigerator. We are in the position of someone who cannot keep the bathtub full because the water keeps running out: before we buy a bigger nuclear- or coal-powered water heater, we really ought to get a plug. Technologists have lately invented some very clever plugs. The new generation of jet aircraft now being manufactured is about twice as fuel-efficient as the present fleet. The best aluminum smelters use about 40 percent less energy per pound than the present average. Compared with the present U.S. fleet average of about 16 miles per gallon, the average imported cars now being sold are twice as efficient (32 mpg); a diesel Rabbit, which has only 10 percent less space inside than the average domestic car made in 1978, gets nearly three times as much mileage (45 mpg); its turbocharged version gets four times as much (60–65 mpg); and Volkswagen has already tested an advanced Rabbit with a city/highway rating of five times as much (ca. 80/100 mpg). Still better cars—even big, comfortable ones—have been made experimentally, and 100 mpg is straightforward with present technology.

Consider how many BTUs (British Thermal Units) of heat it takes to warm a square foot of floor space in your home through one Fahrenheit degree-day of cold weather outside. A typical "sieve," such as most of us live in, requires about twenty BTUs of heat. Most recently built sieves need ten to fifteen. What the federal government until very recently regarded as a well-insulated house would need about eight or ten. But Gene Leger, an engineer in East Pepperell, Massachusetts, recently built a tight, heavily insulated house that costs and looks the same as any other house on the block but uses only one and one-third BTUs. The best present art for "super-insulated" houses in Saskatchewan (Canada) and elsewhere uses about a quarter of a BTU or less.

The best new office buildings use less than a tenth as much energy as normal ones, look the same, are more comfortable to work in, and repay from energy savings any extra construction cost within the first few years.

In fact we now know how to make cost-effective houses or other buildings which require essentially no heating in a subarctic climate and no cooling in a tropical climate. Such houses are being routinely built by contractors in Canada and California. They are heated by the warmth gained from people, windows, lights, and appliances, and are cooled mainly by not getting hot in the first place. They are very well ventilated by way of heat exchangers (which you can build for about $80) that recover 80 percent of the outgoing heat or "coolth," so you do not have the stuffiness, or indoor air pollution, otherwise associated with a tight building. Further, it is now possible to bring many existing buildings nearly up to these impressive standards of energy efficiency, using techniques developed in Canada and Sweden.

These ways of saving energy are presently available, presently economic (cheaper than going out and getting more energy, and often even cheaper than buying the energy we are now using), and have no significant effect on lifestyles. They wring more work from the energy we already have, rather than requiring us to cut back on the services that energy provides. They mean efficiency, not curtailment. These two ways of saving energy are not the same—insulating your roof does not mean freezing in the dark—though they are often confused.

The Price Is Right

Detailed studies in over a dozen industrial countries, including the United States, have shown that if we supply energy services in the cheapest way—by using more productively the energy we have—we can increase our standard of living while using much *less* total energy (and electricity) than we do now. For example, Roger Sant, federal energy conservation manager under the Ford administration, has shown how much energy Americans would have bought in 1978 if, for about ten years before that, we had "minimized consumer costs through competition," simply by making the energy investments to provide desired services at least cost. In this "least-cost strategy"* we would have bought about 28 percent less oil (cutting imports at least in half), 34 percent less coal (avoiding stripmining the West), and 43 percent less electricity (rendering unnecessary over a third of today's power stations, including essentially all the nuclear ones). The bottom line: We would have paid about *17 percent less money* for the same energy services we actually received. The savings would be far larger compared with today's energy prices, or with the even higher prices we would pay for newly ordered power stations and synthetic fuel (synfuel) plants. Sant's

*Summarized in *Harvard Business Review,* May–June 1980, pp. 6 ff.

most recent research even suggests that a least-cost strategy over the next twenty years would *decrease* total energy costs as a fraction of GNP. Far from driving inflation, the energy sector would then become a net exporter of capital to the rest of the economy.

Indeed a detailed study* commissioned in 1979 by then Deputy Secretary of Energy John Sawhill has shown that a least-cost energy strategy during 1980–2000, using only technologies already operating in the U.S., could reduce total energy use by at least a quarter, and the use of nonrenewable fuels by nearly half, even if the size of the economy simultaneously increases by two-thirds and there are no significant changes in how we live or how we run our economy. Such a strategy would also provide surplus electrical capacity even if, by the year 2000, all oil- and gas-fired, nuclear, and old power stations had been retired.

The energy industry concedes that energy-saving measures are possible and economically worthwhile, but claims that their development is so slow that we will need more power plants in the meantime. Surprise! The saving we have gotten from better efficiency is already the fastest growing part of our total energy supplies. In 1979, 98 percent of our economic growth was fueled by energy savings, and only 2 percent by all the energy supply expansions (such as increased coal and nuclear capacity) put together. Millions of individual market choices by housewives, industrialists, truckers, farmers, and many others to save money by saving energy have provided the energy basis for more economic growth than the centrally planned supply programs— by over fifty to one. By 1980–1981 this ratio had risen to more than one hundred to one. Even earlier, during 1973–1978, we already *got* twice as much new energy "supply" from efficiency improvements, twice as fast, as synthetic fuel advocates claim they can do at about ten times the cost. Despite more than $100 billion in annual tax and price subsidies which make conventional fuels and power look cheaper than they really are, energy prices are now high enough to give people a strong economic incentive to raise their "energy productivity."

Further, because such energy-saving measures as insulating your home take days, weeks, or months to install rather than the ten years it takes to build a power plant, they are now our fastest growing source of energy. Efficiency measures are much simpler to build, and more accessible to far more people, than the large supply systems. The problems they pose are far more tractable than those common to such gigantic, complicated, controversial energy systems as nuclear power.

Thus the answer to the question of how much energy we will need seems clearly to be "much less than previously believed." Both the empirical evidence of what has been happening and thorough analysis of what could happen if we pursued a least-cost energy policy show that it is unnecessary to debate whether coal is nicer than nuclear—because we will need neither.

A New Prosperity: Building a Sustainable Energy Future (Andover, Mass.: Brick House, 1981).

Technology Is the Answer! (But What Was the Question?)

The old view of the energy problem embodied a worse mistake than forgetting to ask how much energy we needed: it sought more energy, in any form, from any source, at any price—as if all kinds of energy were alike. This is like saying "All kinds of food are alike; we're running short of potatoes and turnips and cheese, but that's okay because we can substitute sirloin steak and oysters Rockefeller." Some of us have to be more discriminating than that. Just as there are different kinds of foods, so there are various forms of energy whose different prices and qualities suit them to different uses. There is no "demand for energy" as such: nobody wants raw kilowatt-hours or barrels of sticky black goo. People instead want energy services: comfort, light, mobility, ability to bake bread or make cement. We ought therefore to ask "What are the many different tasks we want energy *for,* and what is the *amount, type, and source of energy that will do each task in the cheapest way?"*

Electricity is a special, high-quality form of energy. It can do difficult kinds of work efficiently, and it is correspondingly *very expensive.* An average kilowatt-hour delivered in the U.S. in late 1980 was priced at about 5¢, equivalent to buying the heat content of oil costing $80 per barrel. A nuclear power station ordered in 1980 will deliver electricity costing (in 1980 dollars) at least 8¢ per kilowatt-hour, equivalent on a heat basis to about $130 per barrel—four times the 1980 OPEC oil price! Lights, motors, electronics, appliances, subways, and smelters all require electricity and may be able to justify its high price. The other 92 percent of our energy needs are for heat (58 percent) and vehicular liquid fuels (34 percent). But such premium uses, only 8 percent of all delivered energy needs, are already met twice over by today's power stations.* Electricity used for water and space heating, for air conditioning, and for nonrail vehicles can never give value for money, no matter how efficient the heating/cooling device or electric propulsion system, because there are far cheaper ways to do these same tasks.

Thus generating more electricity is an irrational response to the energy problem we actually have. Arguing about what kind of new power station to build—coal, nuclear, solar—is like shopping for the best buy in brandy to burn in your car, or the best buy in Chippendales to burn in your stove.

In fact, *any* kind of new power station is so uneconomic that if you have just built one, you will save the country money by writing it off and never operating it. Why? The additional electricity it might generate could *only* be used for low-temperature heating and cooling, because the special "electricity-specific" needs are already fully met. But low-temperature heating and cooling can be provided much more cheaply by weatherstripping, insulation, heat exchangers, greenhouses, window shades and shutters, window overhangs and coatings, trees, etc. These measures generally cost

*The electricity now being generated and sold is about two-thirds more than can be used to economic advantage (not counting the scope for further efficiency improvements). The generating capacity available for use (but not fully used) is about twice the worthwhile level.

about 0.4¢ per kilowatt-hour, whereas the running costs *alone* for a new nuclear plant will be nearly 2¢ per kilowatt-hour,* so it is cheaper not to run it. Under our crazy U.S. tax laws, the extra saving from not having to pay the plant's future subsidies and profits is probably so big that writing it off would also recover the sunk capital cost of building the plant!

But suppose that even though the nation has more electricity than it can cost-effectively use, a particular region needs additional power, say, to meet the needs of a growing population. That need should be supplied from the cheapest sources, and all sources should be forced to compete on their economic merits. In approximate order of increasing price, sources of additional electricity are:

1. Eliminate pure waste of electricity, such as lighting empty offices at headache level.
2. Save the electricity now used for space conditioning and water heating by substituting good architecture (heat-saving measures, passive solar heating and cooling, and then—if needed—carefully designed active solar heating). Some U.S. utilities now give zero-interest weatherization loans, which you need not repay for ten years or until you sell your house—because it saves them millions of dollars to get electricity by saving it compared with building new power plants.
3. Make appliances, motors, lights, smelters, etc., cost-effectively efficient: the parable of the refrigerator.

Just these three measures could quadruple U.S. electrical efficiency: we could then run today's economy, with no changes in lifestyle, using only a quarter as much electricity, just by using it in an economically efficient way. Such a small electrical demand could be met without needing *any* thermal power stations, old or new, fueled by oil, gas, coal, or uranium, using only present hydroelectric capacity, readily available small-scale hydro, and a modest amount of windpower.

Should still more electricity be needed, however, the next cheapest sources would include:

4. Industrial co-generation, combined-heat-and-power stations, low-temperature heat engines run by industrial waste heat or by solar ponds, upgrading existing big dams to full capacity, modern wind machines or small-scale hydro turbines in good sites, and perhaps even presently commercial solar cells using waste-heat recovery and optical concentrators.

It is only after we had clearly exhausted all these cheaper opportunities that we would even consider

5. Building a central power plant of any kind—because that is the slowest and costliest known way to get more electricity or to save oil.

* About 1¢ (in 1980 dollars) for fuel-cycle costs, 0.5¢ for operating and maintaining the plant, and 0.6¢–0.7¢ for operating and maintaining its associated extra transmission and distribution capacity. Long-term waste management, decommissioning, reserve margin, regulatory and security costs, and externalities (public and occupational risks) are omitted here.

Getting Off Oil

Given the wide array of measures, especially the diverse ways of saving energy, that can reduce our dependence on foreign oil and make our energy supplies sustainable, we need to compare different oil-saving measures to be sure we are using the cheapest, fastest, surest package of measures. Every dollar devoted to relatively slow and costly power plants actually *slows down* oil displacement because it cannot also be spent on more effective measures. This is certainly the case with nuclear power. After thirty years and over $40 billion in direct government subsidies—nearly $200 for every woman, man, and child in this country—nuclear power is still providing less than 2 percent of our delivered energy needs. Two percent is smaller than just the *growth* in delivered energy from private and industrial woodburning that has occurred in the past five years through individual initiative with virtually no subsidies. When was the last time someone called for a tenfold expansion of firewood facilities to meet our energy needs?

The only realistic way to reduce our oil dependence quickly is through efficiency improvements: stop living in sieves and stop driving Petropigs. The sieves are obvious enough: most of us live in houses that have holes in the walls totaling a square yard, little insulation, and no heat exchangers to recover heat escaping in the outgoing air and water. Basic weatherization of homes and commercial buildings during the 1980s could save at least 2.5 million barrels of oil per day, at an average price of about $6–$7 per barrel. That saving is two-fifths of the 1980 rate of net U.S. oil imports (6.2 million barrels per day), or about two-thirds of the much lower 1982 rate (under 4 million barrels per day).

We could save more than our entire net oil imports by one other relatively simple measure: replacing gas-guzzlers with very efficient cars. Gas-guzzlers have such a low trade-in value that they are filtering down to poor people who can least afford to run or replace them. This keeps inefficient cars on the road longer, when we should be getting rid of them faster. Rather than building synfuel plants (or reactors), it would be cheaper and faster to save oil by using the same money to *give* people a 40-mpg car—provided they will *scrap their Brontomobiles* to get them off the road. (No trade-ins, because then someone else will drive them and defeat the whole purpose: the cars should be recycled and a death certificate provided.) Alternatively, for every mile per gallon by which your new car improves on your old car which you scrap, it would be worth giving you a $200 cash grant. (That would pay back in an average of five years, as compared with synfuels at $40 per retail barrel—about half the likely price.) Gas-guzzlers scrapped and not replaced should also attract a bounty.

The economic benefits of speeding up the scrapping of gas-guzzlers would be enormous even without a subsidy. Suppose that Detroit switched in one giant leapfrog straight to making 50-mpg cars and light trucks, saving about 4.5 million barrels of oil per day. Suppose that the extra retooling cost—beyond the $50 billion they're planning to spend on retooling in the 1980s anyway—were as implausibly high as $100 billion, about enough to rebuild Detroit from the ground up. Even that high figure, spread over a complete new fleet of cars and light trucks, would pay back in *sixteen months* against the 1980 gasoline price of $1.25 per gallon!

Thus just the two biggest oil-saving measures, pursued just in the 1980s, to a level well short of what is technically feasible or economically worthwhile, could together save more than the entire 1980 net rate of oil imports—before a synfuel plant or a nuclear or coal-fired power station ordered in 1980 could deliver any energy whatsoever, and at about a tenth of its cost! If we do not pursue such cost-effective measures, it is hard to take seriously our alleged concern about reliance on foreign oil—especially if instead we spend a trillion dollars (as the utility industry proposes) on new power plants, even though the oil used by old ones is only 7 percent of all the oil we burn, is falling fast, and could be eliminated sooner by simply fixing up buildings.

So What Is the Energy Problem?

Most energy planners have been trying to solve the wrong problem. Consider, for example, a sad little story from France, involving a "spaghetti chart" (stylized in Fig. 6–4), a device used by energy planners to show how our energy flows from primary sources through various conversion processes (the "meatballs" in the middle) into final forms that are put to various end uses. (Losses along the way are so high with electrification—because power plants turn about three units of heat into only one unit of electricity—that according to Department of Energy forecasts over two-thirds of the primary energy growth during 1980–2000 will be lost before it ever gets to consumers.)

In the mid-1970s the energy efficiency planners in the French government started, wisely, with end uses—the right-hand side of the spaghetti chart. Their biggest single need for energy was to heat buildings. They calculated that even with good heat pumps, the most uneconomical way to heat buildings would be with more electricity. They fought and won a battle with Électricité de France, their nationalized utility; so electric heating was to be discouraged or even phased out as a waste of money and fuel.

Meanwhile down the street, the energy supply planners, who were far more numerous and influential in the French government, started on the left-hand side of the same spaghetti chart. They said: "Look at all that nasty foreign oil coming into our country. We must replace that oil. Oil is energy," they mused; "we obviously need some other source of energy. Voilà! Reactors give us energy; we'll build a lot a nuclear reactors!" They paid little attention, however, to what that extra electricity could be used for, or to its price.

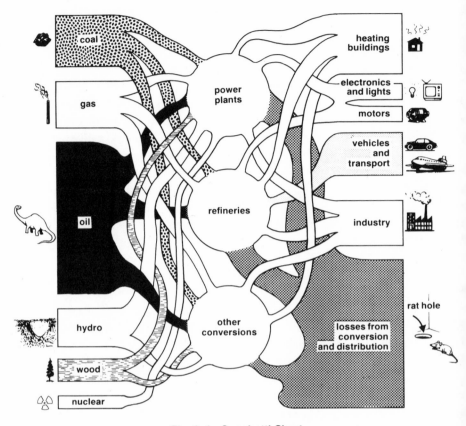

Fig. 6–4   Spaghetti Chart.

Thus the two sides of the French energy establishment went on with their respective solutions to two different, indeed contradictory, French energy problems: *more energy of any kind,* versus *the right kind to do each task in the cheapest way.* In 1975 these conflicting perceptions collided as the supply planners suddenly realized that the only way they would be able to *sell* all that nuclear electricity would be for electric heating, which they had just agreed not to do!

Every industrial nation is in this embarrassing position of supplying more electricity than it can use cost-effectively. If one starts on the left side of the spaghetti chart arguing that the problem is a shortage of energy, it is easy to see nuclear power—or any other energy supply technology—as the answer. It is only by starting on the right side of the chart that one can begin to match the task to be done with the technology which can supply that energy service (whether via greater efficiency or increased supply) at least cost.

Soft Technologies

Efficiency improvements can indeed squeeze a great deal more usable energy services out of our present system, but eventually additional, sustainable supplies of energy will be needed both to meet new demand and to replace old technologies. What are the cheapest, fastest, surest, safest sources of the amounts and kinds of energy we need?

The answer is widely assumed to be the centralized, "hard" technologies, like synfuel plants, power plants, Arctic oil and gas pipelines, liquefied natural gas (LNG), and perhaps in the next century solar power towers and solar satellites. But in the past few years conservative economic analyses in many countries, based on real cost and performance data, have shown a surprising turnabout from this conventional answer. If we shop carefully, the cheapest buys, after efficiency improvements, turn out to be the "soft technologies"—the many diverse renewable sources (running on sun, wind, and farm and forestry wastes) that are relatively understandable to the user (though, like a pocket calculator or digital watch, they can be technically very sophisticated) and that supply energy of the right *scale* and *quality* to do each task in the cheapest way. After the soft technologies, as the Harvard Business School's 1979 energy study found, the *next* cheapest investments are the synfuel plants; and last, most expensive of all, are new power stations. We have as a nation been taking these options in reverse order—worst buys first.

Soft technologies are not necessarily cheap, but they are much cheaper than not having them. They are consistently cheaper in capital cost, cheaper still in working capital needed, and several times cheaper in energy price to the consumer than are the synfuel plants and coal or nuclear power stations which would *otherwise* be needed to do the same tasks. As we have documented elsewhere,* the soft technologies are also faster, yielding more energy, money, and jobs back sooner per dollar invested. They are better than the hard energy alternatives in other ways too—for jobs, the economy, the environment, Third World development, protecting the climate, stopping the spread of nuclear bombs, preserving our freedoms and our reliance on individual and local choice, and making our energy supplies less vulnerable to disruption.†

All the soft technologies we have included in our analysis are now in or entering commercial service. They include the best present art in:

> low-temperature passive solar heating and cooling, and active solar heating;

> high-temperature solar heating for industry (we now have technologies to do this even on a cloudy winter day in Scandinavia);

> converting farm and forestry wastes (but not special crops) efficiently into liquid fuels for efficient vehicles, taking care to preserve soil fertility;

* See *Energy/War: Breaking the Nuclear Link,* footnoted previously.
† See A. B. Lovins and L. H. Lovins, *Brittle Power: Energy Stratgy for National Security* (Andover, Mass.: Brick House, 1982), which also documents the technological and economic status of soft technologies.

○   present and small-scale hydroelectricity; and

○   a modest amount of windpower for electrical generation, heat pumping, water pumping, and mechanical and hydraulic drive.

Some utilities talk about power towers in the desert, monocultural biomass plantations, multimegawatt wind machines, ocean-thermal-electric conversion, wave power, or solar power satellites, and say that "Solar power isn't here yet." But we do not assume any of those technologies. We do not even assume cheap photovoltaics (solar cells), even though the Department of Energy now expects that solar-cell technologies already developed will become competitive on a large scale by 1986.

Are the soft technologies enough to meet the long-run energy needs of a dynamic economy? We and many of our colleagues abroad have carefully analyzed this question. Conservatively assuming only present technologies, competitively priced, properly designed, and efficiently used (unlike many official studies which have only demonstrated incompetent engineering—it *is* possible to do solar wrong), we have all found that present soft technologies are enough to meet virtually all long-term energy needs in every country so far studied. These include not only the United States and Canada, but also such countries as Britain, France, West Germany, Denmark, Sweden, and Japan—countries that are, in various combinations, cold, cloudy, northern, heavily industrialized, and densely populated. The mix of sources varies widely from one country to another, and between different parts of the same country, but there always seems to be enough to meet each nation's energy needs.*

Though most soft technologies are relatively small and (like the energy flows they tap) dispersed rather than centralized, that does not result from some dogmatic notion that everything has to be small. It would be just as silly to run a giant smelter with little wind machines as to heat houses with a breeder reactor. The object of appropriate scale is to minimize the costs and losses of energy distribution, just as matching the quality of energy to its task minimizes the costs and losses of energy conversion. Throughout our analysis, when we found soft technologies to be superior to "hard" ones such as nuclear power and synfuels, our criterion was simply that of orthodox market economics: meeting people's energy needs at least cost. We left out all the "external" advantages of soft technologies, such as increased employment, cleanliness, and safety; those merely strengthen our argument. We also assumed that hard technologies could do safely all that their promoters claim, at lower cost than they claim. We assumed many conservatisms that weight our calculations in the sense least favorable to our conclusions. Further, we assumed massive economic and industrial growth (even though that may be unrealistic on other grounds) in order to show that if your goal is to "Los Angelize" the planet, you will be able to meet the resulting energy needs most cheaply and effectively with a soft energy path. If you think such a goal is unworthy, or if you feel that today's values and institutions are somehow imperfect, then you are welcome to assume instead a mixture of technical

*See *Least-Cost Energy: Solving the CO₂ Problem,* footnoted previously.

and social change that would make a soft energy future even easier to implement. But we have not done that. We have analyzed and described a "pure technical fix" in which people's energy needs are met just as conveniently and reliably as they are today, without changing lifestyles.

Let's Do It

As the hard energy path steepens and grows more rocky, the soft path beckons. Greatly increasing our energy productivity, gradually switching in the coming decades to the soft technologies, and meanwhile using fossil fuels cleanly and efficiently for the transition is not only socially and politically attractive—it is all we can afford. Properly done, this path also avoids the political costs of a more centralized, electrified, nuclearized energy future: greater concentration of political and economic power, more inflation and unemployment, inequitable distribution between who gets the energy and who gets the social costs, more bureaucratization, more vulnerability to disruption by accident or malice, and from efforts to counter this vulnerability, more erosion of civil liberties. The soft path also averts the chilling prospect that in a few decades we shall have tens of thousands of bombs' worth of plutonium per year circulating as an item of commerce within the same international community that has never been able to stop the heroin traffic.

The soft path does not require and indeed cannot tolerate central management: it is implemented by individual choice through the existing market and political process; it preserves competition and free enterprise; it is available to rich and poor, rural and urban; it increases our resilience and national security in the face of a surprise-full future. This convergence of political and economic logic is already leading the people of the United States, as individuals and as communities, to implement a soft path with surprising speed. Already in the past four years, Americans have gotten more new energy from renewable sources than from any or all of the nonrenewables, and have ordered more megawatts of small hydro capacity and windpower than of coal or nuclear power stations or both of them put together.

There is much we can do, however, to speed the transition. Desubsidizing the energy system to ensure genuine competition, having energy prices that tell the truth, and purging the obstacles to sensible investment would all smooth the way. But despite such institutional barriers as lack of consumer information, inequitable access to money, obsolete laws and rules (some enacted by utilities to ban clotheslines, wind machines, or co-generation), and the split between landlords and tenants as to who pays for and who benefits from energy savings, known solutions to these problems are entering their early stages of implementation, and astonishing initiatives are starting to bubble up through the cracks.

Why all this activity? Some people are afraid of oil cutoffs. The citizens of Fitchburg, Massachusetts, weatherized over half of their houses in a ten-week door-to-

door action program. Some low-income communities are now busily building sola
collectors in neighborhood factories or erecting community heat and food-producin,
greenhouses because solar energy gives them energy independence at a price the
can afford. Affluent communities like Davis, California, are taking the soft path t
the good life. From the high deserts of New Mexico, where one house in every 30
has installed a solar greenhouse, to energy-efficient Portland, Oregon, to the Frankli
County Energy Project in Massachusetts, now starting to implement its answer t
imported oil with a community program of local insulation, wind, microhydro, an
solar, the evidence of what works is proving that most people can solve their ow
energy problem. All it takes is incentive and opportunity—and the realization th:
this is *our* problem. Made of so many billions of little pieces scattered throughout
big and diverse society, the energy problem cannot be solved by central management
Energy is not too complex or too technical for ordinary people to understand—thoug
it may be too simple and too political for many technical experts to understand.

In short, we are relearning what Lao-tzu said some 2500 years ago:

> Leaders are best when people scarcely know they exist,
> not so good when people obey and acclaim them,
> worst when people despise them.
> Fail to honor people, they fail to honor you.
> But of good leaders who talk little,
> when their work is done, their task fulfilled,
> the people will all say: "We did this ourselves."

A Golden Decade for Solar Energy?

MODESTO A. MAIDIQUE

Modesto A. Maidique is an associate professor of engineering management at Stanford Un
versity. He formerly taught at MIT and at the Harvard Graduate School of Business Admi
istration. He is one of the co-authors of *Energy Future,* the celebrated book of the Energ
Project at the Harvard Business School, and also president and chief executive officer of Co
laborative Research in Lexington, Massachusetts, a biotechnology firm. With expertise in bo
solid-state physics and technology management, Professor Maidique evaluates solar energy
potential and comes up with some surprising conclusions.

The decade of the 1980s can be a window of opportunity for the new energy tec
nologies and in particular for solar energy, that is, for renewables, broadly define
Fossil fuels are no longer cheap and plentiful. Nuclear technology is stalemated. Th
promise of synthetic fuels will not be realized for ten years or more, if at all. Fusic
is decades away. The price of conventional energy continues to rise, albeit wi
misleading interruptions. On the other hand, new "breakthroughs" in solar techno

ogy are announced every week. Generous federal and, in many cases, state solar tax credits have been simultaneously implemented, while costs for many of the solar technologies are dropping. Federal funding of solar research projects exceeded half a billion dollars in 1980.

In ten years the situation may be quite different. The bloom, in fact, may already be coming off the solar rose. The Reagan administration's budget slashing has cut deeply into solar programs. The new budget proposals are less than half those of the Carter days.* Denis Hayes, perhaps the world's best known solar advocate, and one of the most respected, was fired in June 1981 from his post as head of the Solar Energy Research Institute. To Hayes, the Reagan administration has declared "open war on solar energy." A favorite of solar advocates, the Solar and Conservation Bank, has repeatedly failed to be signed into law.

Later in the decade the outlook may be even dimmer. The solar tax credits will probably have been phased out. One or more synthetic fuel technologies might be well on their way to commercialization, putting a long-term lid on the real price of oil and gas. An acceptable nuclear waste disposal technology might be on hand, thus unlocking the handcuffs on nuclear energy. Safer means to mine, transport, and burn coal will have been developed. Environmental standards may even be relaxed from present levels. And appliances, automobiles, and factories will have become far more energy efficient. Solar energy in the late 1980s may look far less attractive than now.† To prevent the shrinking demand for solar because of cheap alternatives, as occurred in Florida and California earlier in the century, solar must make rapid progress in this decade.

This then defines a window of opportunity, a decade during which the preconditions exist for the new wave of energy technologies, including solar, to stake out their position in the U.S. energy portfolio of the future. Whether or not the now-burgeoning field of solar technology can take full advantage of this opportunity is an open question. Many barriers—technological, economic, political, institutional, behavioral, and even ideological—stand in the way of accelerated solar growth.

Solar, broadly defined, encompasses a wide variety of energy technologies at different stages of development, including fuels from plant matter, hydropower, wind, direct solar heating, and solar photovoltaic solar satellites. This essay is concerned primarily with those solar technologies that are here, or close to being here, such as the on-site solar technologies and fuels from plants. These technologies are *here-and-now alternatives to imported oil,* not gleams in the eyes of scientists. On the other hand the research solar technologies, such as ocean thermal power and solar space satellites, will compete in the much more distant and unpredictable energy conditions of the twenty-first century.

However, it can be anticipated that solar's many friends in Congress will restore some of the original funding.
On the other hand accelerating solar technological improvements may more than offset gains made in other technologies such as coal burning.

In this decade of opportunity, what is the solar position? Solar energy already supplies about 6 percent of the nation's energy—about twice the contribution of nuclear energy—primarily in the form of hydropower, wood, and waste. Over a quarter million solar heating units have been installed in the U.S., the bulk of them in California and the Southwest. Although aggregate savings from solar heating are still less than the equivalent of 100,000 barrels of oil per day, the industry is still in an embryonic stage, and the opportunity exists for increasing this contribution by ten times or more from present levels. Meanwhile gasohol stations have sprung up all over the country. Ethyl alcohol production is increasing rapidly, and 10–20 percent of our automobile fuel may be supplied by gasohol (gasohol is 10 percent alcohol) by the end of the decade. The number of wood stoves has increased fivefold, to almost five million during the last five years.

What fraction of our national energy input could this array of new alternate sources contribute by the year 2000? Predictions vary widely, from a few percent to almost half of our energy supply. One selection of high-quality projections* varies from 7 percent to 25 percent (see Table 6–1). The truth, however, is that no one really

TABLE 6–1   PROJECTION OF SOLAR CONTRIBUTION BY THE YEAR 2000

Source of Projection	Projected Solar Energy (in mbdoe)[a]	% of Total
President's Council on Environmental Quality (CEQ)	12	23
Walter Morrow, Lincoln Laboratory	14	16
Stanford Research Institute 1 (Business as Usual)	5	7
Stanford Research Institute 2 (Low Solar Cost)	10	13
Stanford Research Institute 3 (High Fuel Cost)	6	12
National Academy of Sciences (CONAES)	4	8
Union of Concerned Scientists (Maximum Demand)	10	20
Union of Concerned Scientists (Intermediate Demand)	10	25
National Audubon Society	10	25
Solar Energy Research Institute (SERI Draft)	7.5	25

Source: Adapted from M. A. Maidique, in *Energy Future,* edited by R. Stobaugh and D. Yergin (New York: Ballantine Books, 1980), p. 263.
[a]Million barrels per day of oil equivalent.

*Including a Solar Energy Research Institute draft which has become the focus of an acrimonious debate between solar advocates and the Reagan administration because it puts the governmental imprimatur on a prediction that coincides with that of many solar activists.

knows what the solar contribution will be in the year 2000. There are too many intervening economic, political, and technological variables. But it is reasonable to say that if we make a strong commitment to solar energy *now,* by the year 2000 solar's 6 percent contribution could reach about one-fifth of our national energy needs. If solar were to reach this level of contribution by the year 2000, one plausible breakdown among the various solar alternatives is roughly that shown on Table 6–2.

Despite these tantalizing possibilities, many obstacles still stand in the way of solar-powered America. Most forms of on-site solar energy—small windmills, solar heating, and small dams—require that the user assume increased management responsibilities for energy, something that most consumers are often reluctant to do. These technologies frequently require rebuilding, modification, or redesign of existing structures. In short, as one homeowner declared, "Solar energy is a hassle."

Beyond the "hassle," on-site solar energy requires several times more capital per BTU than conventional energy sources, or alternative new technologies such as synthetic fuel plants. And even assuming capital availability, the on-site technologies are often not economically competitive with traditional alternatives, even at present energy costs. Analyses can be prepared, however, that show that small-scale solar energy is actually far more economical than first believed, particularly in view of the continuing sharp increases in conventional fuel costs. But most consumers and many corporations ignore such analyses and look only at simple payback.

There are also powerful institutional barriers such as building codes that do not include solar and laws that do not recognize access to the sun as a right. In time solar energy will have its own environmental opponents, if only on the grounds of esthetics, as has already occurred in Coral Gables, Florida, where the city planning committee came close to banning solar collectors because of their awkward appearance.

Perhaps the most subtle barrier is an ideological one. Solar energy has traditionally been associated with dreamers, idealists, and tinkerers. As such, solar energy has not been perceived by most business executives as a realistic investment oppor-

TABLE 6–2   BREAKDOWN OF SOLAR ENERGY SOURCES IN THE YEAR 2000

Energy Source	Oil Equivalent (million barrels per day)
Solar space and hot-water heating (including active and passive)	2
Other on-site technologies	2
Wood and waste	4
Hydropower (large installation)	2
Total	10

Source: Adapted from M. A. Maidique, in *Energy Future,* edited by R. Stobaugh and D. Yergin (New York: Ballantine Books, 1980), p. 267. Assumes a 50-mbdoe scenario for the U.S.

tunity. Conversely, solar advocates have long viewed the business establishment as a reactionary opponent. As one solar enthusiast explained, "Big business and the utilities are the enemy." Paradoxically, some of the largest investments in solar technology have been made by major firms such as Exxon, Asarco, the Grumman Corporation,* and ARCO. Firms like General Electric and Owens-Illinois have made innovative product entries to the solar collector field.

At least one solar energy technology, fuels from plants, is largely exempt from many of these barriers: capital requirements are comparatively modest, the economics can be excellent, and if used to fire centralized burners or to produce gasohol, no additional managerial burdens are placed on the consumer. Wood, for instance, is perceived by business as an eminently realistic alternative. After all, wood and plant matter once supplied 95 percent of the energy of Colonial America. For these reasons fuels from plants will probably be the single largest solar energy category in the national energy account. By the year 2000 as much as four million barrels per day of oil equivalent may be displaced by plants and their derivatives, the energy equivalent of about half the oil production of Saudi Arabia in 1979. The main barrier to wood usage will be opposition from those who object to using forest land for energy and to the pollution that is produced by burning wood.

But let us return to the problem of diffusing the on-site technologies. As in the case of nuclear energy, a central problem is that these technologies are not yet being marketed as a complete system.

It may stretch the imagination to compare a solar hot-water system to a nuclear energy system. Nonetheless they are both, in principle, systems. And just as the absence of an accepted waste disposal technology and adequate safety measures make nuclear a lame technology, the would-be solar supplier who does not provide a means of capital supply for his capital-hungry solar installation is selling an incomplete system. The buyer does not want simply to purchase a collector that generates heat he wants a proven, guaranteed, complete system that does not significantly change his conventional way of buying energy in monthly installments. To overcome this problem we need large-scale business investments in solar energy. For this to happen we must dispel the notion of solar energy as a romantic, pastoral technology. Those who wish to restructure America's political system radically should not attempt to do it on the anvil of solar energy. They should simply argue for a political revolution Solar energy activists must not deal with illusions, but with realities. If solar energy is to become big business one day, solar heating panels and windmills will have to be made in plants similar to those where tractors, washing machines, and automobiles are manufactured in large volume. There is nothing magical about solar energy; it is simply another energy source which, to succeed, will have to be perceived by business as a major and profitable opportunity.

As emphasized earlier, no one can predict what the future holds for solar energy

---

* Grumman, however, has cut back its solar efforts, following heavy losses in its solar business.

There are many political, technological, economic, and institutional uncertainties. A simple listing of the challenges to be met is enough to discourage the most sanguine analyst, which may explain the skepticism about solar energy. But this kind of analysis overlooks a basic fact: where an economic market and a viable product have existed, our private enterprise system has always found ways to overcome the most resilient obstacles. OPEC is on our side to reach this goal. With every price increase, every supply interruption, solar energy becomes relatively more attractive. A solar energy market boom follows every energy shortage.

We must do two things if we wish to diffuse solar energy on a broad basis: First, we must find innovative ways to make solar energy more competitive. Material costs—primarily aluminum, copper, and steel—account for 80 percent of the cost of a solar collector system. We must find ways of tapping the sun's energy that do not use these materials, or we must build the collectors into the construction of the sites to be heated, or develop cheap photovoltaic systems as substitutes. We need that gush of innovation in solar energy that our inventors and entrepreneurs have traditionally provided at other times in American history. Second, and most important, we must encourage business, especially big business, to complete the solar technology system. To do this we must make solar energy part of the establishment. Solar is at the point where we should thank the dreamers, idealists, and "mother earth types" who helped bring it to where it is, and to begin to identify it as a viable and realistic part of the energy sector establishment. Only then will it be able to lay claim to the resources that it justly deserves, not only from the government but from the private sector as well.

This is already beginning to happen. Recently I spent a day with twenty-five senior executives at a major oil company. The topic: solar energy as an investment opportunity. And I have spoken with countless executives from small, medium, and large firms who are asking the same question: Is this the time to invest in solar energy? What about wood stoves, solar heating, or photovoltaics? This is a very different situation from what prevailed two or three years ago, when businessmen would smile condescendingly when solar energy was put forth as an investment opportunity.

The government can also help. It can accelerate the solar commercialization process in at least three ways: (1) by funding basic research that is too risky or too complex for individual private firms to undertake; (2) by providing tax credits of 30–50 percent to overcome the initially sluggish phase of the innovation cycle; and (3) by purchasing substantial quantities of solar energy equipment for use in its own buildings and installations. The latter may be the most effective of the available mechanisms. Studies of the semiconductor industry indicate that it was government purchases of semiconductor components rather than government-sponsored research grants that was the key stimulus in the industry's early years. Whereas basic research may not have an impact for decades, and tax credits may be partially absorbed by manufacturers, government purchases create a market today that will lead to a broader market tomorrow. This is the kind of signal that industry has traditionally taken seriously.

The next decade and the following one will witness the impact of successive technological waves in the solar energy field. There will be great advances, but there will also be setbacks. Support for publicly funded research programs and incentives will rise and wane. The future outlook will change from year to year. But if Solar America is to become a reality, one thing is clear: private enterprise will lead the way. The same entrepreneurial spirit that paved the way for the automotive revolution and later ushered in the computer era must unlock the door to the solar world.

Closing in on Fusion

STEPHEN O. DEAN

Dean is chief executive officer of Fusion Power Associates and also director of Fusion Energy Development for Science Applications, Inc. He is a consultant for the Lawrence Livermore National Laboratory, the Oak Ridge National Laboratory, and various industrial organizations, and from 1972 to 1979 he was director of the Magnetic Confinement Systems Division for the DOE. Dr. Dean explains why "break-even" is no longer the scientific problem that has haunted fusion research for the past two decades. In fact three countries are now building demonstration (research) fusion machines to test the technological feasibility of fusion in the mid-1980s.

Fusion, as well as fission, can unlock the tremendous energy that binds the nucleus of the atom. But it is fusion—a process eight times more powerful than fission—that generates the heat of the sun and the stars. The sun, which makes life itself possible, and the stars are large natural fusion reactors. Fusion has been described as the "ultimate energy source" in the universe, not only because it "powers" the sun and the stars, but also because fusion fuels are in nearly limitless supply on earth. However, the achievement of practical and economical fusion power is probably the most difficult technological feat yet attempted by man.

Fusion could go a long way toward solving the world's energy problems for several reasons. First, the basic fuel in a fusion reactor is deuterium,* a heavy form of hydrogen, which is readily available to all nations from ordinary water. Second, the *amount* of energy released by the fusion reaction is enormous by conventional standards. Fusion fuel releases a *million* times more energy than does burning a comparable weight of coal or oil; one teaspoon of deuterium, obtained cheaply from a gallon

---

*The difference between hydrogen (H) and deuterium (D) is that the hydrogen nucleus consists of a single proton, whereas the deuterium nucleus contains a proton and a neutron. When H or D are combined with oxygen, they form, respectively, ordinary water (H_2O) or heavy water (D_2O). In nature, one out of every 6500 molecules of ordinary water contains deuterium. The cost of separating the deuterium from a gallon of ordinary water is about ten cents.

of water, contains the energy equivalent of 300 gallons of gasoline; a mere 1000 pounds of deuterium could fuel a 1000-megawatt power station for a year.

Fusion power plants potentially offer a number of environmental and safety advantages over other energy sources, such as fossil-fueled power plants and fission reactors. Compared with fission reactors the absence of such fission products as radioactive iodine and cesium from the fusion fuel cycle reduces the potential hazard of any accident by more than a thousandfold. The waste from a fusion reactor is significantly less toxic than fission waste; tritium,* a radioactive gas which will be used with deuterium to fuel first-generation fusion reactors, is relatively harmless (biologically) and short-lived. Some radioactive waste will be produced in fusion by the activation of materials that surround the fusion process (steel vessels, piping, etc). However, these materials typically will have radioactive half-lives of tens rather than hundreds of years. This short half-life, plus the nonvolatile nature of the radioactivity, suggests that this material can be buried as low-level waste. By experimenting with the use of special types of alloys, such as vanadium rather than stainless steel, engineers should also be able to control the amounts of this radioactivity so that the "waste" can be reused within a few decades. And unlike uranium ore, which must be milled, enriched, or reprocessed, the two fuels in a fusion reactor pose no complicated environmental hazards: deuterium is extracted from sea water, and tritium is produced inside the reactor.

You might ask whether or not an accident similar to the one at Three Mile Island could occur in a fusion reactor. The answer is no. In the core of a nuclear fission reactor, heat from fission reactions is generated in solid fuel rods both when the reactor is operating and for some time after the reactor has been turned off. A continual flow of coolant must pass over these rods to remove the heat; otherwise the temperature of the rods will rise and melting could occur, releasing radioactivity into the coolant system. In a fusion reactor the core is an extremely dilute, albeit very hot, gas. If there were an accident in the containment vessel and the gas were to escape, it would cool down instantly when it touched the solid walls. Even a complete loss-of-coolant accident could not cause the equivalent of a "meltdown" because the amount of fuel gas in the reactor is so small (approximately a ten-thousandth of atmospheric pressure) that the total heat content is insufficient to melt through the vessel walls.

Because fusion power is not yet commercially developed its costs are not yet established. Fusion energy should eventually become economically competitive with other energy sources (e.g., coal, oil, gas, or nuclear fission), because of near-zero fusion fuel costs, compared with coal, oil, and uranium, which will surely escalate.

When will we begin to get electricity for our homes from fusion power? We hope early in the next century. Scientists expect to reach the "break-even" point (where

---

*Tritium, another form of hydrogen, has one proton and two neutrons in the nucleus. Although tritium is not found in nature, it can be produced within the reactor from lithium, a relatively abundant metal found throughout the world.

as much energy is produced by fusion as is needed to heat the fuel to fusion temperatures) by the mid-1980s. Now that the scientific feasibility of fusion is no longer in doubt, scientists are turning their attention to the more practical engineering development of commercial fusion power. The engineering of commercial fusion reactors, which will likely be comparable in size to today's power plants, is expected to be somewhat more complex than today's fission reactors, but much is yet to be learned. In 1980 the U.S. Congress passed Public Law 96–386, which sets as a national goal the operation of a commercial demonstration plant around the turn of the century. The law calls for a doubling of the current $400-million-a-year government fusion budget within seven years, but the Reagan administration's austerity will likely inhibit these plans, at least in the near term.

Two Approaches

Fusion reactions occur under extreme and exacting conditions. The nuclei of deuterium and tritium atoms naturally want to repel each other because both are positively charged. In order to fuse them together scientists must heat the deuterium-tritium gas mixture to such high temperatures that the electrons surrounding each atom become dislodged and the gas turns into a plasma*—a collection of free nuclei and electrons.

*Plasma is the fourth state of matter (the other three are gases, solids, and liquids). The sun and other stars are large balls of plasma. A plasma behaves very similarly to a gas, except that in a plasma the particles are electrically charged (i.e., ionized) rather than neutral.

Fig. 6–5 Fusion: Like the Sun and the Hydrogen Bomb. A. Fusion energy—the process that powers the sun and the hydrogen bomb—is achieved when two atoms of a lighter element (e.g., hydrogen) are joined to form a heavier element (e.g., helium). Scientists are trying to generate controlled fusion power in the laboratory in two ways: by *magnetic confinement,* which simulates the conditions of the sun, and by *inertial confinement,* which applies the principle of the H-bomb. B. In the sun the enormous force generated by fusion tends to blow the sun apart, but is balanced by the sun's own gravitational force, which tends to collapse the sun. These two forces cancel each other out, creating a stable mass of extremely hot plasma. C. On earth, scientists replace the gravitational field by a magnetic field. The expanding force of fusion in a hot plasma (contained in a doughnut-shaped container) is held together by large magnetic forces. D. The hydrogen bomb releases the energy of fusion through a different mechanism. Within the hydrogen bomb the enormous temperatures necessary to fuse hydrogen nuclei are achieved by detonating an atomic bomb (schematically represented as two hemispheres of plutonium-239 which are rapidly blown together by chemical explosives to achieve critical mass). The detonation of the atomic bomb within the H-bomb creates enormous amounts of x-rays and other forms of energy, which are used to compress and heat a mass of hydrogen-rich material (e.g., lithium deuteride). The rapid collapse of this material creates the conditions for fusion. E. Laser fusion operates on the same principle as the hydrogen bomb. Tiny pellets of hydrogen-rich materials are dropped successively into a chamber, where intense laser beams are focused on them with sufficient intensity to crush them. The sudden implosion of these pellets creates the conditions for fusion.

Pairs of deuterium and tritium nuclei can then fuse into single nuclei of the gas helium, ejecting high-speed neutrons in the process. The neutrons, which carry away most of the energy of the fusion reaction, are absorbed by a "blanket" about three feet thick that surrounds the plasma vessel. As the blanket gets hot, a coolant passing through the blanket is heated so that it can produce steam, which can be used to drive a turbine to generate electricity.

Plasma temperatures must remain very high in order to overcome the electric repulsion of the nuclei. Nature uses gravity to hold together the plasmas of the sun and the stars. Scientists on earth can contain plasma in one of two ways: they can

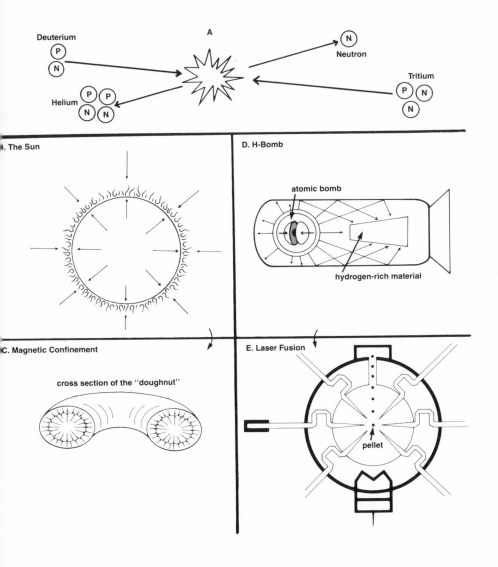

suspend the plasma in magnetic fields or they can use the inertia* of a hot fuel pellet, which has been bombarded and compressed by beams of either laser light or energetic particles (see Fig. 6–5).

Magnetic Confinement

Magnetic fields provide invisible lines of force that can hold a plasma somewhat in the way that a magnet exerts a force on iron fillings. A machine called a tokamak (invented by Soviet researchers) holds plasma in a doughnut-shaped magnetic field, much as an inner tube holds air.

A tokamak currently under construction at the Princeton Plasma Physics Laboratory (the largest fusion energy research center in the world) is expected to be the first magnetic fusion reactor to break even. Scheduled for completion in 1982, the Tokamak Fusion Test Reactor (TFTR) is expected to reach the necessary conditions for break-even: high temperature, sufficiently high density of plasma, and time of confinement (see "Break-Even").

Tokamaks are not the only kind of magnetic container, however. In the magnetic-mirror machine, the plasma is held in a cylindrical container that has strong magnetic fields at each end. The fields reflect, or mirror, the fusion plasma back toward the center when it attempts to leak out through the ends. The Lawrence Livermore National Laboratory in California is conducting most of the research for the magnetic-mirror program and is building a Mirror Fusion Test Facility (MFTF) which will use deuterium fuel, scheduled for completion at Livermore by 1985. So far, small magnetic-mirror devices have heated the plasmas to about 230 million degrees Celsius, but they have not yet reached the required product of density-time. The TFTR and MFTF, however, are major scale-ups in size over earlier experiments and are expected to bridge the gap between laboratory research and large test reactors.

Inertial Confinement

In this process, which uses no magnetic fields, the enormous temperatures and densities are created by the power of lasers or electrically charged particle beams. Relying on principles on which the hydrogen bomb was based, scientists use intense beams of light or atomic particles to bombard a tiny fuel pellet from all directions. The heat of the beams vaporizes the outer layer of the pellet, and the "blowoff" of this outer vapor creates an inward force that compresses the pellet to tremendous densities while also heating it to high temperatures. At a sufficiently compressed density (about *one thousand times* the normal density of solid materials) and temper-

* Inertia is the tendency of matter to remain in place or with a constant velocity unless acted on by outside forces.

In order to generate not only the energy required to heat the deuterium and tritium to a plasma state, but also the additional (net) energy to produce electricity, fusion reactors must be able to satisfy three conditions. First, the plasma must be heated above 50 million degrees Celsius (three to six times the estimated temperature of the sun's interior). Second, large numbers of these nuclei must be concentrated at a high density in order to produce a high rate of energy release. Third, particles in the plasma must be confined at these temperatures and densities for sufficient time to guarantee the release of a useful amount of energy.

The energy released is dependent both on the rate of deuterium and tritium fusions and the length of time over which this rate is maintained. For example, approximately the same amount of energy can be released by a high density of nuclei confined over a short period of time as by a low density of nuclei confined over a longer period of time. The longer the holding time and the greater the density of the fusion nuclei, the larger the amount of energy released in the fusion process. To break even, the temperature of the plasma must exceed a minimum of 50,000,000° Celsius, and the *product* of the density (n) multiplied by the confinement time (t) must exceed 10^{13} seconds per cubic centimeter. For example, this value could correspond to a confinement time of one second and a density of ten trillion nuclei per cubic centimeter, or a confinement time of ten seconds and a density of one trillion nuclei per cubic centimeter. (By way of contrast, the density of air is about ten million trillion molecules per cubic centimeter.)

ature, the deuterium and tritium in the fuel can ignite, fusion can occur, and a huge surge of energy can be released before the pellet expands. This concept relies on the pellet's resistance to change in motion—inertia—which holds the pellet together for a billionth of a second before it blows apart from internal pressure. For a practical power plant, such pellets would have to be ignited approximately ten times per second. Experiments have successfully achieved the ignition temperature at low densities, but are still about ten times short of the requisite compression.

Shiva, a twenty-beam laser device the size of a football field at the Lawrence Livermore National Laboratory in California, is capable of producing nearly 200 terawatts (200 trillion watts) for a billionth of a second. To date, Shiva has compressed fuel pellets to over 100 times the normal density of ordinary liquids, which is 10 percent of the density required for break-even. An upgrade of the Shiva laser, a $200-million fusion machine called Nova, is under construction at Livermore to extend these results to higher compressions.

The research into inertial fusion, conducted at the Los Alamos Scientific Laboratory, Lawrence Livermore National Laboratory, and the Sandia Laboratories, is supported by military funds because of the government's interest in simulating nuclear explosions using these techniques.

It is too early to assess whether the magnetic approach will prove to be more cost-efficient than the inertial approach in the long run. Government officials and most scientists believe that magnetic confinement may be the best bet for the first commercial reactors.

Scientific Progress

The task of developing fusion for practical use began in 1952, after the hydrogen bomb was detonated at Bikini Atoll. Not only was it a scientific challenge, but an engineering one as well. The implications of fusion for future society were judged to be so profound that the program, code-named "Project Sherwood," was cloaked in secrecy for six years. Then in 1958, as part of President Eisenhower's Atoms for Peace program, all nations working on fusion (which included the United States, Britain, and the Soviet Union) declassified their programs and pledged to cooperate to make commercial fusion a reality for the benefit of all.

The painstaking task of developing the basic science and technology required for civilian fusion power continued at a relatively slow pace through the 1960s. In the 1970s the years of dedicated effort began to pay off; success followed success. For example, scientists at MIT working with the Alcator A tokamak achieved adequate confinement in late 1975. In mid-1978 Princeton scientists proved that the minimum ignition temperature could be attained in a tokamak. The three major tokamaks under construction (in the United States, Europe, and Japan) are aimed at achieving fusion break-even conditions in the mid-1980s. Scientists are confident that these experiments will be successful.

In 1980 an advisory panel to Congress stated that "fusion can be made commercial before [the year] 2000 if a national commitment is made soon," and an advisory board to the DOE concluded, "As a result of this [recent scientific] progress, the United States is now ready to embark on the next step: . . . exploration of the engineering feasibility of fusion." Congress called for a Center for Fusion Engineering (CFE) to be established and a Fusion Engineering Device (FED; based on the best available design) to be built and operating by 1990. President Carter signed the Magnetic Fusion Energy Engineering Act, which calls upon the federal government to get a commercial reactor on line near the turn of the century.

Major corporations, always interested in potentially attractive energy technologies, are investing research and development money, although most industrial activity is through direct government contracts. The electric utility companies (primarily through the Electric Power Research Institute) are investing several million dollars per year in fusion research and development in the United States through direct contracts and are providing advice and consultation to DOE program management. A

WHAT ABOUT WASTE?

Unlike cesium, iodine, and other radioactive fission products that must be stored for hundreds of years, the product of a fusion reaction is helium—a stable, harmless gas which is *not* radioactive. There are, however, two kinds of radioactive materials produced by the fusion reaction, and these must be handled safely.

The first is tritium gas, which conceivably could accidentally leak out of storage tanks at the reactor site. In its radioactive decay a tritium nucleus emits a beta particle (an electron) which has so little energy that it cannot even penetrate the outer layer of the skin. Tritium gas mixes well with air or water, so enormous dilutions could be accomplished easily. Tritium gas could be hazardous if it were inhaled, but it would remain in the body for only minutes or seconds. Tritiated water (produced in small amounts in the reactor coolant or in the air) if ingested would remain in the body for about ten days, as compared to years for strontium or cesium. Furthermore, tritium is a million times easier to dilute than iodine and a hundred thousand times easier to dilute than plutonium.

The second kind of radioactive material is found when atoms in the fusion reactor's structural material absorb the neutrons emitted from the fusion process and become radioactive themselves (this is called neutron activation). The neutrons not only make the surrounding metal mildly radioactive, but also weaken the metal by causing minute structural dislocations in the metal. Portions of the structure would have to be changed periodically and disposed of, but because this metal is not volatile, the radioactivity cannot escape into the environment like a gas or liquid. Also, the total amount of such radioactivity can be greatly reduced if construction materials that become less radioactive are used.

Possible problems with fusion waste, however, cannot be pinpointed because a commercial fusion power plant has yet to be developed. Perhaps by 2020, when fusion power could be competitive with other energy sources, scientists will have developed exotic deuterium-deuterium reactions, which contain much less radioactive tritium.

---

major government-funded fusion program (approximately $35 million per year) exists at General Atomic Company, and the team of Ebasco Services and Grumman Aircraft Corporation is building the TFTR. Industry and the utilities will gain more incentive to invest as the program draws closer to commercialization.

"We're not making any money on fusion," said Ted Stern, executive vice-president of nuclear engineering systems at Westinghouse. But the company "feels that fusion energy might be a winner."

A Promising Future

After nearly three decades of developing the fundamental scientific principles of fusion energy, scientists are beginning to devote their attention to engineering obstacles that must be overcome. For example, scientists hope to develop such advanced (nontritium) fusion reactions as deuterium-deuterium, that could eliminate almost all the radioactive by-products of the fusion process. Until then substantial inventories of tritium will have to be safely managed (see "What About Waste?").

Industry must play an increasingly larger role in fusion development, laying the groundwork for more accurate economic assessments of fusion potential, if fusion is to become commercial shortly after the turn of the century. While fusion energy will probably not be cheaper than other energy forms in the early twenty-first century, we expect it to become increasingly competitive as time goes by.

Perhaps the most important reason to support the development of fusion is that it could unlock a limitless fuel supply from water, with no geographical boundaries, and thus reduce world tensions caused by the ownership and manipulation of scarce fossil or nuclear fuels. Fusion has appeal for many scientists and engineers, especially the young, because of its unique challenges: new scientific technological frontiers combined with its potential major impact on future society. Commercial fusion power, though still many years away, is coming within our grasp.

Breeder Reactors: The Next Generation

ROBERT AVERY AND HANS A. BETHE

Robert Avery is a theoretical physicist who has led numerous research teams to analyze accidents in light-water and breeder reactors. Since 1973 he has been director of the Reactor Analysis and Safety Division of the Argonne National Laboratory near Chicago. Hans A. Bethe was director of the theoretical physics division of the Manhattan Project in Los Alamos during World War II. He has made major scientific breakthroughs in a wide variety of fields: theory of atoms, collision theory, radiation of electrons, theory of the solid state, shockwave theory, nuclear physics, astrophysics, and quantum field theory. He is a professor emeritus of physics at Cornell University. In 1967 Dr. Bethe was awarded the Nobel Prize for determining the complex sequence of nuclear processes that generate the power of the sun and the stars. In this essay Dr. Bethe and Dr. Avery argue that the benefits of inexhaustible nuclear energy from the breeder far outweigh the breeder's potential drawbacks—proliferation of plutonium and criticality accidents—which, as they explain, are manageable.

The energy crisis has made it clear that current energy sources cannot be used indefinitely. Oil, gas, coal, and even uranium are running out, becoming too expensive, or are inadequate to handle our future energy demands.

However, there is another energy source, the breeder reactor, which promises us virtually unlimited energy. The breeder will extend our limited uranium reserves, which now are estimated to last perhaps several more decades, to 100,000 years and beyond. Scientists and engineers are not the only ones impressed with this promise of inexhaustible energy: scores of nations (including every major industrial one) have chosen the breeder to provide the major part of their electricity in the next century.

The commercial light-water reactors (LWR) of today, which dominate the reactor market in the U.S. and around the world, are wasteful of uranium. They can only consume a rare form of uranium, U-235, which comprises a tiny fraction (0.7 percent) of all natural uranium. By contrast, the breeder consumes essentially all of the uranium, especially U-238, which makes up 99.3 percent of all natural uranium. Quite simply, breeders hold the promise of unlimited energy by converting "useless" U-238 (which is plentiful but difficult to fission) into plutonium-239 (Pu-239), which is fissionable. In this way waste uranium is turned into a valuable fuel.

We will describe the breeder reactor, explain why it is essential if nuclear power is going to meet our future energy needs, and present a discussion of the perceived hazards, real or imagined, of this technology. We will focus on the breeder type that dominates the scene today, the fast breeder reactor operating on the uranium-plutonium cycle.

Why Is the Breeder Needed?

It is generally agreed in the scientific community that among the long-range inexhaustible energy options the fast breeder has by far the highest technical, engineering, and economic feasibility. While significant work still remains to be completed in methodically scaling up fast breeder technology to meet commercial standards, no major scientific or economic developments, such as those required for solar electric and fusion power, are necessary.

The reasons that the fast breeder has achieved this unique status are straightforward. The fast breeder is a natural extension of the present reactor technology and could readily be built by the existing nuclear industry. Moreover, because the breeder fuel cycle extends the efficiency of uranium utilization by a factor of at least 100, it is feasible to mine a much lower grade of ore than is used for present reactors. (There is much more total uranium available in these lower grade ores than in the higher grades, and therefore the supply of fuel is extended much more than a hundredfold.) It is also likely, but not yet fully proven, that uranium may be recoverable from sea water or from granite. This would represent an effectively infinite supply of fuel for the breeder.

Additional benefits accrue to the breeder from the following two factors. Hundreds of thousands of tons of uranium that has already been mined and processed exists above ground and is available as the raw material for the breeder program. This uranium has had most of its "fissile" isotope, U-235, removed and is therefore no longer useful for light-water reactors. But it provides excellent raw material for breeders; there are hundreds of thousands of tons of this uranium available, so no fresh uranium would have to be mined for many hundreds of years. Also, a very large amount of plutonium, in the form of spent fuel, is potentially available from light-water reactors. This discharged fuel would be essentially wasted if there were no breeder program.

An incidental benefit of a breeder program is that it simplifies the disposal of nuclear waste. The present U.S. program would bury the entire spent fuel elements from LWRs, including the plutonium they contain. Because this plutonium has a long half-life, over 24,000 years, some groups have expressed concern over this long-lived radioactivity. Although we do not share this concern because the nuclear waste will be buried safely and deeply, the concern would be removed if the plutonium were burned in breeder reactors.

How Breeders Work

In one sense breeders are like commercial light-water reactors. Both are fission reactors, operating on chain reactions caused by the fissioning of U-235 or Pu-239. Like the light-water reactor, the breeder reactor has a core of fissionable material, ten to fifteen feet in diameter, consisting of several hundred subassemblies. Each subassembly consists of several hundred vertical pins less than a half inch in diameter. Inside the pins are the uranium and plutonium fuel (see Fig. 6–6).

The difference between the breeder and the LWR arises because breeders produce more "fast" neutrons than the light-water reactor (see "How Breeder Reactors Differ from Light-Water Reactors"). In both the breeder and LWR the neutrons released by the fissioning of the U-235 (or Pu-239) nucleus travel at enormous velocities (thousands of miles per second). These fast neutrons do not fission the uranium nucleus well, so in a commercial LWR they are deliberately slowed down to low velocities (one mile per second or less) because slow neutrons (called *thermal neutrons*) fission the uranium nucleus more readily.

Scientists in the 1940s solved the problem of slowing down the neutrons from the chain reaction by surrounding the fuel with water (or carbon). The collisions between the fast neutrons emerging from the fissioning process and the hydrogen nuclei found in ordinary water slow down the neutrons, making them more suitable for sustaining the chain reaction. Thus the water surrounding the fuel of a LWR actually serves *two* purposes: it both carries away the heat energy of the core and decreases the velocities of the fast neutrons.

The breeder, by contrast, requires fast neutrons. When fast neutrons fission the plutonium nucleus, they create more excess neutrons than slow (thermal) neutrons do when they are absorbed in either uranium or plutonium. These neutrons may then be captured by U-238, and when this occurs the uranium is converted to plutonium-239. But one must see to it that the neutrons fission plutonium while they are still

In a breeder reactor, as in any other nuclear reactor, on the average one of the neutrons released in fission must cause a new fission in the next generation in order to continue the nuclear chain reaction. Any excess neutrons may be captured by "fertile" nuclei (e.g., U-238) to create "fissile" nuclei (e.g., Pu-239). In principle when more than two neutrons are released per fission process, any remaining neutrons (above the one required for the next fission) can create more than one new fissile nucleus by absorption in a fertile nucleus.

In practice, however, considerably more than two—usually three or more— neutrons have to be produced because of the inevitable losses that will occur in the system due to parasitic capture (the absorption of neutrons by elements other than uranium or plutonium, such as fission products or structural materials) and neutron leakage out of the system. If the amount of plutonium produced exceeds the amount consumed, the system is defined to be a breeder.*

Even though more fissile material is being produced in the breeder than is being consumed, the fuel in the system must be recycled, that is, periodically chemically reprocessed and refabricated, in order to remove the fission products that gradually bring in increasing amounts of parasitic absorption.

Because of the low concentration of fissile nuclei in a LWR, slow neutrons are essential to maintaining the chain reaction. The hydrogen nucleus, having approximately the same mass as the neutron, is able to slow down the neutron by collisions with it in the same way that a billiard ball hitting another of equal mass may give up a large fraction of its energy to the second ball.

Although in a breeder reactor fast neutrons do not fission the Pu-239 nucleus as readily as do slow neutrons, fast neutrons have the advantage that they produce more neutrons per fission, yielding more excess neutrons which make breeding plutonium possible. Furthermore, fast neutrons are not readily captured by the nuclei of fission products in the core and they produce some additional fissions in the fertile U-238. For these reasons fast neutrons are needed to maintain the large excess number of neutrons in the core required for breeding plutonium.

In a fast breeder reactor it is necessary to exclude all materials of low atomic weight which might slow down the neutrons. Little neutron energy is lost in collisions of the neutrons with the materials of high atomic weight (now analogous to a billiard ball hitting an object of much greater mass and bouncing off with very little loss of energy). Therefore in the fast breeder system a coolant having a relatively high mass nucleus, usually sodium, or extremely low density, usually helium, is necessary.

---

*The breeding or conversion ratio is the ratio of the amount produced to the amount consumed (e.g., a typical breeding ratio might be 1.2, which means a 20 percent net gain in fissionable fuel). By contrast, a typical LWR might produce 0.6 new fissile atoms per fissile atom consumed from fission, and therefore there is a net gradual loss of fissile material.

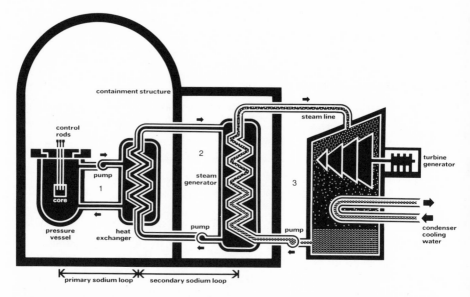

Fig. 6–6   Liquid Metal Fast Breeder Reactor (LMFBR). A LMFBR has three cooling loops. Molten sodium coolant in the primary loop is pumped through the reactor core containing the fuel. This sodium carries away the heat of the core and transfers it to a secondary sodium loop via the heat exchanger. The second loop transfers the heat to the steam generator, which turns the water in the third loop into steam. The turbine blades are set in motion by the steam, creating electricity by the process of induction. [Courtesy Atomic Industrial Forum, Inc.]

fast, before they are slowed down by collisions with other nuclei, such as U-238 or the sodium coolant; therefore, breeders exclude very light materials and also use fuel that contains a lot of plutonium, 10–20 percent rather than 3 percent U-235 in the LWR fuel.

In a breeder the core containing the U-238 and Pu-239 fuel is surrounded by a "blanket" of uranium-238. It is the purpose of the blanket to capture escaping neutrons from the core to produce plutonium-239. The fertile U-238 in the blanket is gradually converted into Pu-239. The blanket is removed from the reactor and shipped to a reprocessing plant where the blanket is chemically treated to extract the Pu-239 from the waste fission products and the U-238. The Pu-239 can then be recycled back into the same or into other breeder reactors.

The breeder coolant, sodium, flows in between the pins at a temperature of about 600°F and leaves the core at about 900°F. The primary sodium, that is, the sodium that cools the core, then goes to the intermediate heat exchangers where the energy— but essentially none of the radioactivity—is transferred to an intermediate sodium loop. The intermediate sodium travels to a steam generator/superheater system which produces the steam that goes to the turbine generator system. The steam produced in the breeder is considerably higher in temperature and pressure, and therefore the breeder is more efficient than the LWR in converting heat into electricity. This is because sodium remains liquid up to 1600°F at atmospheric pressure, while even at very high pressures water can only reach much lower temperatures. Thus another fundamental difference between the two systems is that the LWR is kept at a very high pressure in order to get suitable steam conditions. (For example, the PWR is kept at approximately 2000 pounds per square inch, the BWR at approximately 1000 pounds per square inch.) In contrast, the sodium-cooled fast reactor operates at, or slightly above, atmospheric pressure.

As with ordinary commercial reactors, a number of different kinds of breeders exist. We describe the one that has become dominant, the sodium-cooled mixed plutonium–uranium oxide fast breeder, referred to as the Liquid Metal Fast Breeder Reactor (LMFBR). There are other fast breeder types, based on different coolants (such as helium), different fuels (such as mixed uranium and plutonium carbides or mixed uranium and plutonium in metal form), and different fuel cycles (such as the thorium-uranium cycle, where thorium is converted into fissile uranium-233, a rare isotope of uranium).

There is one major variation in the fast breeder design: the pool versus the loop system. In the pool concept the entire primary system (the core, primary pumps, and intermediate heat exchangers) is submerged in a large sodium-filled tank. With everything contained within the tank the need is eliminated for an intricate system of pipes connecting the primary system components. In the loop system the intermediate heat exchangers and the primary pumps, rather than being submerged in a large tank along with the core, are located within separate, steel-lined concrete cells. A large amount of external piping containing the primary radioactive sodium is required to connect the primary system components. This disadvantage of the loop design may be offset by the easier accessibility for maintenance. A debate exists as to which of these two systems is superior. So far both systems have been about equally developed, even though there does seem to be a trend, in the larger systems, toward the pool system.

Safety

Nuclear opponents claim that the breeder technology presents a number of undue hazards. However, intensive studies of the breeder have concluded that breeder accidents present an extremely low risk, probably even lower than for the light-water systems, and far below other risks to which the public is ordinarily exposed.

A major reason that breeders may be safer than LWRs is that the liquid sodium coolant is kept at ordinary atmospheric pressure, which reduces the risk of coolant leakage. Furthermore, the sodium is far below its boiling point of 1600°F so that there is a large margin to high pressure. Moreover, sodium conducts heat very effectively which tends to reduce local hot spots in the reactor. The potential disadvantage of liquid sodium is that if it leaks, it reacts chemically with water and burns in air.

However, the chances of leaks are reduced by the lower pressures found in breeders.

The primary concern with the breeder reactor is that it operates at much higher enrichment levels (10–20 percent) than do light-water reactors (3 percent), which somewhat increases the chances that a partial melting of fuel could lead to a runaway chain reaction (the criticality accident). However, contrary to what some critics say, breeder reactors cannot explode like atomic bombs. Many years ago it was proven that even under the most unrealistic and pessimistic assumptions, a runaway chain reaction in a breeder (called "energetic disassembly") would result at worst in an accident comparable to an ordinary chemical explosion, rather than a nuclear explosion. Since that time considerable research with more realistic assumptions has shown that an energetic disassembly accident would release even less energy than previously thought.

How could a disassembly accident occur? Under the most pessimistic assumptions, with all overlapping safety systems failing, a disassembly accident could occur if some malfunction (for example, accidental control rod removal or loss of pumping power) triggered a runaway chain reaction. Neutron levels would rise and the core would heat up. A portion of the core could melt and collapse inward under gravity, which might further accelerate the chain reaction. It is also possible, though unlikely, that an interaction between the molten fuel and the sodium coolant might cause the core to be suddenly compressed, thereby increasing the chain reaction. In all cases, however, elaborate theoretical studies and experiments have shown successfully that there is no tendency for the fuel to compact under gravity or interactions with sodium. But even under assumptions of such compactions, analyses and experiments show that the explosive force of a disassembly accident could be contained within the vessel, and surely within the walls of the reinforced concrete containment building.

In addition there exists a further margin of safety, an inherent safety feature that automatically cuts off a runaway chain reaction: the Doppler effect. As the core heats up in an accident, the rate of fissioning of plutonium-239 actually decreases. Because a larger fraction of the neutrons are captured in uranium-238, higher temperatures make it more unlikely that the nuclei will fission. This serves to self-limit any rapid increase in the chain reaction.

Weapons Proliferation and the "Plutonium Economy"

The Carter administration characterized the breeder reactor as an option of last resort. President Carter's primary argument against the breeder was that the reprocessing of spent fuel and the handling of plutonium would result in the availability of weapons material to hostile nations, groups, or even terrorists. We do not believe that in the United States this is a serious concern. Methods for the secure transportation of fissile material have been developed for many years by the Sandia Laboratory, and are by now extremely safe, both against road accidents and against attack by hijackers.

It is true that in the breeder we would use a great deal of plutonium, and this plutonium must be safeguarded. However, plutonium is easy to safeguard while it is in the reactor because, for one thing, nobody can touch it. Once it has been chemically reprocessed, there are several technical means to make the theft of plutonium very difficult. For example, the United States and Britain jointly have developed chemical processes which leave a lot of radioactive fission products in the plutonium, so that the hijackers would be exposed to lethal radiation if they tried to build a nuclear weapon in their "garage." Only a full-fledged "hot" chemical laboratory would do, and this could easily be discovered.

One should realize, too, that the chemical reprocessing plants which will be needed to reprocess fuel will be highly localized and relatively few in number. One such plant can easily take care of fifty reactors. Therefore with a small guard force—some ten to twenty people—and with the addition of some electronic gadgets, it is perfectly possible to protect such a plant. There is absolutely no reason to fear that the guarding of plutonium in this country would require a large police force and lead to any loss of civil rights.

More important, we should realize that any nation determined to make an atomic bomb would surely choose a more direct route. The basic driving force of nuclear proliferation is motivation, since the knowledge to build nuclear weapons exists and will certainly not vanish even if all nuclear power activities were to cease. Thus the real question is political: Will the international tensions driven by energy shortages and the related economic problems enhance the motivation for proliferation more than the elimination of nuclear power might reduce the opportunity for proliferation? We believe the former alternative is true, but whatever the answer may be, the industrialized nations are pursuing the breeder option vigorously. Clearly the U.S. could exert more influence on the development of institutional barriers to inhibit proliferation by having its own strong program rather than not having one at all.

When Is the Breeder Needed?

Much of the debate on the breeder has centered on the question of when it is required. Clearly that issue is linked to the question of available uranium resources and also the question of the rate of development of nuclear power—both of which have large degrees of uncertainty. The amounts of proven and probable uranium reserves are relatively limited and can only supply a limited number of current generation reactors through their lifetimes. Whether the number of such reactors is, say, 500 or 2000 is subject to debate. However one looks at the worldwide picture of reactor development, it is clear that sometime within the next twenty to fifty years commitments to light-water reactors will exhaust the proven and probable uranium reserves. Thus conservative but realistic estimates of the uranium availability, and the assumption that nuclear power will move ahead reasonably aggressively, leads to the conclusion that development of the breeder may already be late.

The breeder could not be available for wide-scale deployment until early in the next century. Several decades are required to deploy any major new energy-supply system to the extent that its impact is significant. This can be seen in transitions from wood to coal, coal to oil and gas, and the start of the nuclear transition. In addition to establishing engineering and economic feasibility, large-scale industrial capability, and major capital investments, confidence on the part of industry, government, and the public is required. For the breeder, engineering feasibility has already been established and the industrial base is largely in place. What remains is the demonstration that this industry can design, build, and obtain a license for a large breeder power plant. Also, the utilities must acquire experience on which to assess the reliability and economics of the breeder.

We believe the argument is strong for the expeditious development of the breeder. Even if it should come in slightly before fully required for wide-scale deployment, there are tremendous engineering and economic benefits that would accrue if, prior to wide-scale deployment, there has been an opportunity for additional breeder technology development.

Economics

While not enough is known about the precise economics of the breeder reactor, it is now clear that the current generation of light-water reactors in many parts of the U.S., and in nearly all other countries of the world, provide a cheaper alternative to coal-fired, and certainly much cheaper than oil-fired power stations. The capital costs for breeder reactors are frequently assumed to be some factor larger than for light-water reactors. Numbers generally are estimated in the range of 1.3 : 1 for this ratio, even though there are many who believe there is no fundamental reason why breeders should ultimately remain more expensive than light-water reactors. Because the breeder is a net fuel producer rather than a fuel consumer, the fuel cycle costs are certainly less for the breeder than they are for the light-water reactor. Adding both capital costs and fuel cycle costs, the breeder becomes progressively less expensive than light-water reactors as the cost of uranium increases. Therefore for any assumed ratio of capital costs, one can define a market price for uranium at which the breeder penetrates the light-water reactor market. These estimates vary, but typically are in the range of $130 per pound of uranium (uranium is presently $35–$40 per pound). Of course there is always the question of whether the uranium will in fact be available at *any* price. There is no doubt that breeders will become economically competitive with light-water reactors within a few decades.

U.S. Program

Since the inception of the nuclear power program some thirty years ago the U.S. has always had a strong breeder program and a succession of progressively larger demonstration reactors. These include the Experimental Breeder Reactor #2 (EBR-II) in Idaho Falls, the Fast Flux Test Facility (FFTF) in Hanford, Washington, and the Clinch River Breeder Reactor (CRBR) in Oak Ridge, Tennessee.

EBR-II, which is small (20 megawatts of electric power), was completed in 1964 and has operated extremely successfully ever since. Except for size, it has many features that are common to full-scale commercial power plants. FFTF, completed in 1980, will be used primarily for the testing of fast breeder reactor fuel. It is considerably larger than EBR-II but still ten times smaller than a commercial power plant, with many features common to full-scale power plants. Construction on CRBR has not yet started, although many of the long-lead-time components have been ordered and delivered. CRBR is about one-fourth the size of a commercial unit.

As a result of the policy redirections during the Carter administration, the future of CRBR remains unsettled at this time. There is some division of view even among breeder proponents, with some arguing for bypassing CRBR and moving directly to an even larger plant, and thus closer to a 1000-megawatt commercial-size reactor.

But quite apart from the setbacks to the breeder program resulting from the Carter administration policy, the U.S. program, which was preeminent in the world in the mid-1960s, has lagged since that time. Most breeder proponents attribute this primarily to the failure aggressively to initiate and pursue the early construction of prototype reactors, say, in the range of 200- to 1000-megawatt plants. The base technology program in the U.S. remains strong.

In any event, the world lead in breeders has passed to France, and perhaps the USSR, though somewhat less is known about their program (see Table 6–3). Fortunately the Reagan administration strongly supports both the nuclear and the fast breeder program. But it remains to be seen whether Reagan's policies will revitalize the U.S. breeder program.

Conclusion

Every major industrial country in the world has opted for the breeder program as its major next-generation system for the production of electrical power. France, the USSR, West Germany, the United Kingdom, Japan, and except for the interim during the Carter administration, the U.S. have all made the breeder reactor their first choice. Furthermore the energy situation of some of these countries, e.g., France, is such that they perceive their need for the breeder as being far more urgent than does the U.S.

France has established as a matter of national policy the full-speed development

TABLE 6–3   FAST BREEDERS AROUND THE WORLD

Name	Country	Type	Power Megawatts (thermal)	Megawatts (electric)	Date Opera
CONSTRUCTION COMPLETE					
DFR	United Kingdom	Loop	60	14	19
EBR-II	United States	Pool	62.5	20	19
Rapsodie	France	Loop	40		19
BOR-60	Soviet Union	Loop	60	12	19
BN-350	Soviet Union	Loop	1000	350	19
BR-10	Soviet Union	Loop	10		19
Phénix	France	Pool	567	250	19
PFR	United Kingdom	Pool	600	250	19
Joyo	Japan	Loop	100		19
KNK-2	West Germany	Loop	58	20	19
BN-600	Soviet Union	Pool	1470	600	19
FFTF	United States	Loop	400		19
UNDER CONSTRUCTION					
PEC	Italy	Loop	135		—
Madras FBTR	India	Loop	40	15	—
SNR-300	West Germany	Loop	762	327	—
Super-Phénix	France	Pool	2900	1200	—
CRBR	United States	Loop	975	350	—
IN DESIGN					
Monju	Japan	Loop	714	300	—
CFR-1	United Kingdom	Pool	3100	1300	—
BN-1600	Soviet Union	Pool		1600	—
Saône	France	Pool		1200–1800	—

of nuclear power and the breeder so that they can remove their dependence on foreign oil imports at the earliest possible date.

The case for the breeder as the first choice in the U.S. energy program for elec tricity has been clear to nuclear planners since the inception of the nuclear powe program some thirty years ago. Every evaluation has invariably reached the conclu sion that while it is possible that some other energy system, such as solar or fusion might ultimately turn out to be better, each of them faces sufficient scientific, engi neering, or economic uncertainty, and extremely long and uncertain developmen programs, that the breeder is certainly the most logical choice. The only remainin requirement with the breeder is actually going through the necessary steps of bringin it in as a large-scale engineering and economic device suitable for wide-scale deploy ment.

We believe that the fast breeder reactor is essential for meeting future energ needs. It represents by far the best solution to meeting central-station electricity needs with hazards that are minimal when compared to the benefits.

7

Where Do We Go From Here?

Introduction

In 1981 Israeli jets bombed a research nuclear reactor in Baghdad, Iraq. Israel defended the raid, calling it an act of "national self-defense" because Iraq, according to Israel, was planning to divert nuclear fuel from the facility to make bombs. The Arabs charged that the act amounted to little more than nuclear blackmail because the Israelis already had developed a reactor program and enough bomb-grade plutonium to make several atomic bombs. The incident not only dramatically highlighted the intense infighting and huge oil stakes involved in the Middle East, but also underscored one political consequence of developing nuclear technology throughout the world—the proliferation of nuclear weapons technology. In the history of mankind probably no single technology has been so controversial or so vital to the age as nuclear power. This is because nuclear power touches the two fundamental issues that dominate this decade: the question of war and peace, and the energy crisis.

Close analysis indicates that these two issues are tightly interwoven into the very fabric of international relations. For example, it is now generally accepted that an acute shortage of energy supplies and the ensuing scramble that would emerge for Middle East oil would increase the chances of war. This reason alone has led many policy makers to believe that an abundance of energy, nuclear energy in particular, would ease international tensions created by tight energy supplies.

But it could also be argued that the presence of nuclear technology around the world, especially the presence of reprocessing centers, would cause the proliferation of plutonium into unstable regions of the world. The likelihood that minor skirmishes might escalate into major confrontations will inevitably increase as weapons-grade plutonium becomes available to scores of nations.

Whether we continue to develop the technology could well be determined by some of these global factors. At times political considerations clearly override any prudent business decisions of where and how to market the technology. After India exploded a nuclear bomb in 1974, even the nuclear industry conceded that any determined nation with sufficient technological expertise can fabricate a nuclear device from commercial spent fuel. Although safeguards can be introduced to deter the theft of reprocessed fuel (e.g., spiking the plutonium with intensively radioactive substances), ultimately the problem of the proliferation of nuclear weapons is a political, not necessarily technical, issue.

Similarly, on the domestic front the final verdict on nuclear power may be decided more by political considerations, such as public opinion, than by any unbiased technical appraisal of nuclear power. For example, even if it could be proven objectively that nuclear reactors are much safer than other energy technologies, if the public does not perceive the benefits to be greater than the risks, political opposition to the technology could eventually bring down the industry—witness how public opposition to nuclear disposal sites around the country has effectively curtailed, for the present any attempts to initiate a long-term waste disposal program. Even if rigorous safeguards and intensive research can show that a meltdown is virtually impossible or that nuclear waste can be isolated for hundreds of thousands of years, in the final analysis it is the public's perceptions of these risks that may well determine the fate of nuclear power.

And so in the first part of this chapter we have asked two distinguished spokesmen to analyze the broad social and political ramifications of nuclear technology, both at home and abroad. Given our knowledge and progress thus far with the nuclear program, where do we go from here?

According to Alvin M. Weinberg, former director of the Oak Ridge National Laboratory, all the technical problems involving nuclear reactors can be solved—he even offers a compelling proposal for the location of reactors in the future. Our ability to proceed with nuclear power will depend on public attitudes toward the technology; if people cannot learn to live with the risks, Dr. Weinberg doubts that commercial nuclear power will survive past this century. In this chapter Dr. Weinberg explains why he believes that the consequences of not developing the next generation of nuclear reactors—that of the breeder reactor, with an inexhaustible amount of energy derived from plutonium—would be much worse than the few surmountable problems attending increased nuclear power development.

Richard Falk, professor of international law at Princeton University, takes a more pessimistic view of the future role of nuclear power. In Dr. Falk's opinion nuclear power and nuclear weapons are irrevocably wed. Because the bomb has already proliferated to several nations, Dr. Falk believes that any attempts to control the proliferation of nuclear weapons will be exceedingly difficult. Shortcuts to bomb building can be taken by terrorists or any of a score of nations expected to have commercial nuclear power by the year 2000. To Dr. Falk, the choice is clear because the stakes are so high: either proceed with light-water and breeder reactor development, which would accelerate the number of nuclear triggers, or flatly renounce all nuclear energy—commercial and military—and begin steps toward complete denuclearization.

Basing their conclusions on the same information, Dr. Weinberg and Dr. Falk have quite different visions of how we should proceed with nuclear power. But even more is at stake than the future of the technology in the nuclear debate—our lives are going to be affected by which energy options we choose to pursue and how vigorously we pursue them. Just as John Gofman believed that trespass by pollution infringed on his individual rights, so too did Bernard Cohen comment that living without nough energy would cause more hardships for him than living with the few disadvantages he feels accompany nuclear power. As with any technology, because nuclear power cannot be made "perfectly" safe a political process must be determined by which people can decide whether the risks outweigh the benefits.

How we arrive at such a process is a very complex issue, incorporating divergent political opinions about individual rights, industrial progress, and the meaning of democracy, as much as it does technical considerations such as the adequacy of the ECCS or adherence to radiation standards. A commonly heard complaint is that those who oppose nuclear power are using the issue as a vehicle to fight economic growth and capitalism across the board. It is impossible to generalize, for it is becoming increasingly evident that not all critics of nuclear power call for the technology's demise.

But there is a kernel of truth to the opinion that nuclear power has become a "trojan horse" for some activists to promote social and political change in our society. This is apparent, for example, when activists state that they would be opposed to nuclear power even it it could be demonstrated to be safe, clean, and efficient. Nuclear power does not seem to be the issue at stake as much as what it seems to represent to many antinuclear activists: corporate wealth and power. And so we close this chapter by exploring perhaps one of the most decisive issues of the nuclear controversy: Is there a hidden agenda? We have asked David Dellinger, a veteran activist, and Bertram Wolfe, a vice-president of General Electric, to define what they believe to be the motives of those who protest, as well as those who support, nuclear power.

The Future of Nuclear Energy

ALVIN M. WEINBERG

Alvin M. Weinberg is often referred to as the "grandfather of nuclear power." For forty years he has been at the center of nuclear energy development, as a former president of the American Nuclear Society, as a former director of the Oak Ridge National Laboratory, as a member of the National Academy of Sciences, and presently as director of the Institute for Energy Analysis at the Oak Ridge Associated Universities. With the end of the First Nuclear Era in sight in several countries, Dr. Weinberg suggests the need for a few technical and organizational improvements in the nuclear enterprise, and asks that we allow the Second Nuclear Era—that of inexhaustible energy from the breeder—to emerge.

In many ways nuclear energy is a fantastic success: a completely new source of energy is now providing, or is soon scheduled to provide, almost 10 percent of all the energy man now produces. This energy will come from approximately 525 large reactors in thirty-six countries. These reactors, if replaced by oil-fired power plants, would require about ten million barrels of oil per day, which is about one-seventh of all the oil produced in the world. Were the output of these plants used for electric heating, in principle five million barrels of oil per heating day could be displaced; if used to recharge electric vehicles, perhaps six million barrels.

Despite this extraordinary accomplishment the First Nuclear Era seems to be coming to an end in many countries. Will there be a Second Nuclear Era, where nuclear energy occupies a secure niche as a large and permanent source of energy? Or will it simply be an ephemeral bridge to a fission-free future based on the sun, geothermal energy, fusion, and fossil fuels (or at least as long as fossil fuels last or until they are prohibited because of their effect on the climate).

It is impossible to generalize: in Austria the First Nuclear Era has ended already or more accurately, was not even allowed to start; in Sweden a majority voted to end it in twenty-five years; in the United States, some states have temporarily banned nuclear energy. By contrast, in France, Japan, and the Soviet Union, nuclear energy continues to grow rapidly, and plans are going forward for the Second Nuclear Era based on breeders or other high-gain reactors, such as the liquid metal fast breeder or the heavy-water reactor.

The most plausible futures require nuclear energy. A world of eight billion people is almost surely going to demand more energy than we use today, assuming the energy can be found. Ralph M. Rotty of the Institute for Energy Analysis and Wolf Haefele of the International Institute for Applied Systems Analysis visualize a world that uses three to four times as much energy in 2030 as we use now. Were most of this to come from coal, the world would have to mine 25 billion tons of coal each year. I would imagine that the dangers of nuclear energy would pale by comparison.

Yet a nuclear future of this magnitude is also formidable. Even if only one-half of this energy were produced by nuclear reactors, we would be speaking of a world of 7500 large reactors. Is this credible? Even if the world allows the first era, based on light-water reactors and limited by the amount of relatively cheap uranium, will it allow a Second Nuclear Era, based on breeder reactors that can be supplied indefinitely with plutonium fuel?

The scenarios are uncertain. At the Energy and Climate Conference in Munster Amory Lovins argued that improved energy efficiency could assure the amenities we now enjoy for eight billion people, using no more energy than is used now. This amounts to reducing the expenditure of energy per capita from two kilowatts per person to one kilowatt per person. I shall not be here in 2050 when this happy situation is expected to take place, so I shall never know whether or not Lovins will be proved right. Given the uncertainties, to curtail the Second Nuclear Era now on the grounds that the world can live in relative peace with an expenditure of one kilowatt per person, is extremely questionable. My own view is that in planning the future we do best to prepare for a higher rather than a lower energy demand. Nor can we count on the other energy options: each is beset with difficulties familiar to all of us. Nevertheless no one can prove that nuclear fission is here to stay: our responsibility as nuclear technologists is to perfect the fission system so that it remains an available politically acceptable option. Ultimately the future of nuclear energy is a political and economic question whose resolution we nuclear technologists can only contribute to, not decide.

First and Second Nuclear Eras

Since a pressurized-water reactor over its lifetime requires about 6000 tons of uranium, we have always understood that the First Nuclear Era was bound to end, because uranium supplies are limited. How soon we run out of uranium also depends on how much uranium can be mined at a competitive price. What the ultimate usable price of uranium might be in current reactors is set by the price of energy from competitive sources. If the competition is, for example, solar power towers, I suspect that the upper limit for the price of uranium is far greater than we now imagine (though the world, paying so much for primary energy, would thereby be a far poorer place). If the competing source is the breeder, the upper limit might be, say, $180 per pound. This is based on the breeder's eventually costing $500 per kilowatt more than the nonbreeder.)

The reactors and institutions required to manage the nuclear enterprise are already in place; scientists are therefore limited in what they can do to ensure that the First Nuclear Era will run its originally contemplated course without being aborted. Two exceptions to this should be noted. The first is waste disposal: a vigorous, clear demonstration of actual disposal of high-level wastes would probably be as important as any single action to incline the public toward support of nuclear energy. Second, Three Mile Island may have proved that the China Syndrome is a myth: a melt-through with a large release of radioactivity may be physically impossible in many loss-of-cooling accidents. Since 1960 the entire nuclear community has believed that failure of the ECCS would in all cases cause the containment to fail. If this is wrong, we must reexamine many basic assumptions. These considerations, coupled with the observation that most of the iodine was retained in the water that flooded the containment building, represent the best of the good news from Three Mile Island. Nevertheless even in the short run technical modifications can and are being made.

Will the First Nuclear Era be successful enough to evolve into the Second Nuclear Era, where we have perhaps ten times as many reactors as we have now, many of these being breeders? No one can tell. Nevertheless the experience of commercial air travel should teach us two things: first, that accidents tend to diminish as experience is gained, and second, that the public does not perceive risk as simply the product of probability times consequence. The public accepts many small airplane crashes, but reacts much more violently to a few very large crashes, although the total casualties may be the same in the two instances. I would guess that this reaction is at least in part attributable to television: each of us can identify with, and be scared out of our wits by, a large accident that we see in detail on the television screen. The public understands consequences; it does not understand probabilities.

The probability of the incident at Three Mile Island is, according to Rasmussen once in 2500 reactor-years, with his estimates going as low as 250 or as high as 25,000 reactor-years. If the First Nuclear Era amounted to 30,000 reactor-years, and the probability remains once in 10,000 years, there would be about ten Three Mile Islands over the next thirty to fifty years, with the range lying between one hundred and one. This I would judge to be intolerable, not merely because the public would lose confidence in nuclear energy long before the tenth Three Mile Island, but because utility executives, whether private or public, would have lost confidence in nuclear energy. Considering that a nuclear accident anywhere is a nuclear accident everywhere, if nuclear energy is to survive the first era we must reduce the accident probability of Class Nine accidents* by a factor on the order of ten to one hundred, so that at most there would be very few, say, one or two Three Mile Islands, within the First Nuclear Era.

If people are to live with fission over the long term we must reduce the probability of accidents by a combination of technical and organizational improvements in the nuclear enterprise. Perhaps the easiest are the technical improvements. For example, German light-water reactors have two relief valves in series so that if one fails to close, the other is available. Certainly the deficiencies that led to Three Mile Island will be corrected: control panels will be provided with positive indication of valve position.

But beyond the technical fixes, various institutional changes are needed. A small group of us oldtimers in the nuclear business (that is, the now rather elderly group of people who were responsible for setting the enterprise along its present course) convened at the Institute for Energy Analysis in Oak Ridge in 1980 to discuss whether the current lull in reactor orders in the United States might be used to advantage to establish criteria for reactors to meet in the Second Nuclear Era.

One proposal for reducing accident probabilities is to confine nuclear reactors to a relatively few, permanent sites. The advantages of clustered, permanent sites seem compelling to me. They include a larger cadre of experts available on site; better organizational memory, and therefore better operation; more effective security and easier control of fissile material; easier handling of low-level wastes and spent fuel elements on site; easier surveillance of decommissioned reactors; and a surrounding population that understands radiation and is prepared to respond in case of accident such as at existing nuclear sites like Oak Ridge. Beyond this I would imagine that the entities that operate large clustered sites are likely to be stronger than are entities operating, say, a single reactor. For example, elements of the Three Mile Island sequence occurred at the Davis-Besse reactor in Ohio, the Oconee reactor in South Carolina, and at Rancho Seco in California before the entire sequence occurred at Three Mile Island. Had all four reactors been located together, I cannot imagine that Three Mile Island would have occurred. The word would have gotten around about the ambiguity in determining water level after a small loss-of-coolant accident.

*Accidents are categorized on a scale of one to nine, nine being the worst.

The trend toward consolidating siting is unmistakable: of the 525 reactors, 170 (representing one-half of the world's nuclear power outside the United States) are now on sites with four or more reactors. If this trend continues, could we not contemplate a world of 5000 reactors confined, say, to no more than 500 or 1000 sites? I am much more comfortable contemplating such a Second Nuclear Era than I am with one in which thousands of reactors are scattered among large numbers of organizations and sites, and in which the learning rate is correspondingly slower.

The Need for Nuclear

If nuclear energy can be made acceptable, then we must ask another question: Is nuclear power necessary? Obviously the necessity for fission in the next twenty years depends on the availability of alternative fuels or sources of energy, and on the future demand for energy.

Throughout the world 400 million tons of oil are burned each year in central electric power stations; this represents 13 percent of the world's consumption of oil in the year 1978. Should all the oil-fired plants be replaced, say, by 1990, with nuclear plants, the pressure on oil would be reduced significantly.

In recent years it has become fashionable to fault this line of argument as simplistic. Rather than replace oil in electric power stations by coal or uranium, we are asked to believe that we can so reduce our energy demand as to make many existing, let alone future, electric power plants superfluous. In any case the use of electricity for such purposes as heating of houses or of water is inelegant and wasteful and ought to be discouraged.

It goes without saying that conservation must be central to any energy policy—indeed much has already been accomplished. For example, D. Reister of the Institute for Energy Analysis points out that in the United States the ratio of energy to Gross National Product decreased by 10 percent between 1970 and 1978. But as Harvard professors Stobaugh and Yergin point out in *Energy Future,* conservation requires decisions by innumerable consumers; by contrast, increased supply requires far fewer decisions. Thus the prediction of how much conservation is actually achieved, as contrasted to how much is theoretically achievable, is intrinsically more uncertain than is the prediction of how much supply can be increased.

The current mood of rejection of electricity seems to me to be irrational. If oil is scarce and coal and uranium are abundant, it makes sense to replace oil with coal and uranium—even if in so doing one must resort to electric heating or other devices that use electricity. After all, we are not driven by thermodynamic imperatives, such as conservation and efficiency; economic or political considerations (such as reducing our dependence on foreign oil) certainly take precedence over the much-discussed, but often-irrelevant stricture to improve the second law efficiency.

Two technical developments, the electric car and the heat pump, could swing the balance toward an electrical future dominated by large central stations. The recent announcement by General Motors of a car powered by a zinc battery that has a lifetime of 50,000 kilometers, a cruising speed of 80 kilometers per hour, and a range between recharges of 165 kilometers, could alter our attitude toward the electric future. In addition the electrically driven heat pump is a proven device. In the United States 560,000 heat pumps were installed last year, and the number has been increasing by 40 percent each year.

If one concedes that (a) a predominantly electric future is at least as plausible as a nonelectric one, (b) oil will continue to be scarce, (c) the solar technologies will not penetrate on a large scale (cost, intermittency, storage?), and (d) coal is not generally available except in countries that possess indigenous deposits,* then it seems inescapable that fission is necessary, at least in some parts of the world. Beyond this, if the cost of coal per joule of energy reaches that of oil—i.e., if the current $80 per ton of coal reaches $140 per ton with oil at $40 per barrel—then even though the cost of a nuclear reactor is high (say, $1500 per kilowatt), electricity from current reactors probably will still be cheaper than electricity from coal-fired stations.

What about the long term, perhaps 50–100 years from now, when oil and gas have become scarce? Fission, if it survives, eventually would be based on breeders (though not necessarily fast breeders). Speaking of the alternatives, I must admit to being agnostic; fusion and solar *may* turn out to be technically and economically feasible, but there is no way of knowing whether this will be the case, and the uncertainties remain. It is much too early to accept such figures as fifty cents per peak watt, which was the aim of the Department of Energy for solar cells, as more than hope. Prudence requires us to develop the breeder, and to deploy it if the alternatives prove too costly or turn out to be unfeasible.

Much of Western society seems to be afflicted by an environmental hypochondria that undermines and debilitates every massive technology. Is it possible that this hypochondria will pass, and that the public reaction to nuclear energy will eventually be commensurate with its true risks?

I see two possibilities. The first is that we will eventually be heeded in our insistence that nuclear risks must be judged in comparison to other risks. I am optimistic enough to hope that people will eventually place risks of nuclear energy in perspective.

The other possibility is that the estimates of the amount of cancer caused by low levels of radiation could prove to be greatly exaggerated. The large number of later cancers supposedly caused by the worst Class Nine accident occur mostly among a very large number of people exposed to less than 3 millirems per day. If low-level radiation could be shown to be *much* less harmful than is suggested by the usual

*I.e., the United States would hardly be prepared in fifty years to mine ten billion tons of coal, most of which is shipped overseas.

linear hypothesis, then the spectre of a reactor accident's causing hundreds of thousands of casualties would be exorcised.

Three recent findings bear on this all-important issue. First, several scientists have shown that, at least for bone tumors caused by radium, there is a practical threshold—that is, the latent period for appearance of the tumor exceeds the life span if the dose is lower than 39 millirems per day. Also, the third BEIR report of the National Academy of Sciences no longer accepts linearity below 10 rems—yet most of the 45,000 latent cancers estimated by Rasmussen from the worst Class Nine accident are attributable to lifetime doses less than 10 rems.

A second possible misconception is the alleged sensitivity of the fetus to prenatal radiation, supposedly established about twenty years ago by Drs. Stewart and Kneale. It was this alleged sensitivity of the fetus that prompted Governor Thornburgh's order to evacuate pregnant women at Three Mile Island. Yet during the past year a flaw was found in the Stewart-Kneale analysis: namely, that the controls which are the basis for any scientific experiment did not match the cancer cases in many essential respects. This casts doubt on the Stewart-Kneale study.

Finally, I call your attention to the recent findings of Dr. John Totter of the Institute for Energy Analysis which show that mortality from cancer seems to be independent of a country's state of industrialization, and therefore of its level of man-made pollution. One must therefore seek the primary carcinogens not among man-made agents but rather among all-pervasive, "normal" components of the environment. The culprit suggested by Totter is oxygen. His main argument is that normal metabolism of food can convert the oxygen we breathe into a carcinogenic ion (O_2^-). The equivalent dose of radiation from the ion might be between 7 and 30 rems per year; it is *this* flood of radiomimetric ions that, in Totter's view, is an underlying—perhaps the most important—cause of cancer. If one accept's Totter's view, then the lifetime dose of 7 rems of background radiation, even on the linear hypothesis, would account for about 0.3–1 percent of cancer.

Witches, Floods, and Wonder Drugs

It is too early to say how Totter's revolutionary theory about the origin of cancer will be received by the scientific community, but perhaps it will result in the realization that low-level radiation is *far* less damaging than even the linear hypothesis suggests, and that therefore most of the fears concerning the lingering effects of Class Nine accidents, or for that matter of conceivable contamination from leaks from waste repositories, are unfounded. These speculations, if proved correct, should help bring the Western world to its senses with respect to the hazards of low levels of radiation.

I close by drawing from Dr. William Clark's perceptive paper on "Witches, Floods and Wonder Drugs," in which he likens the current environmental hysteria to the fear of witches that swept over much of Western Europe and America in the sixteenth and seventeenth centuries. The symptoms were much like those we now see every night on television: vague discomforts, cattle dying, babies deformed because of industrial miasmas. Consider the hysterical fear exhibited by Middletowners when the NRC proposed to vent 60,000 curies of krypton gas from Three Mile Island: the maximum beta skin dose per person would have been 11 millirem, the whole-body gamma dose 0.2 millirem (compared to Totter's estimate of 7000–30,000 millirems per year from the carcinogenic ion).

Witch hunting flourished for two centuries, especially since it was in the interests of the witch-hunting profession to find and burn witches. It was not until 1610 that the chief inquisitor, Alonzo Salazar y Frias, became suspicious about the alleged connection between witchery and human ills. He ordered an investigation and dis covered that, although more than 500,000 "bona fide" witches had been burned at the stake in the past century, nothing else had changed: people still got sick and died wars and pestilence abounded, crops sometimes failed. Though he did not outlaw witch hunting, he forbade the use of torture to extract confessions: the result was that witch burning, and then witch hunting, fell precipitously.

I do not wish to leave the impression that a Class Nine accident is as innocuous as witches have turned out to be; we know that the mid-lethal dose is 400 rems of radiation, and that in the worst conceivable accident some deaths would occur. But we also know that most of the presumptive casualties and the fear of Class Nine accidents comes from low-level exposure. Whether or not there will be a Second Nuclear Era will depend on the public's ability to overcome its unreasoning dread of our modern witch: exposure to low-level radiation. I would hope we will lay to rest this modern witch soon enough to ensure that the First Nuclear Era runs its course and the Second Nuclear Era be allowed to coexist with the solar or fusion era.

Denuclearization*

RICHARD FALK

Richard Falk is the Albert G. Milbank Professor of International Law and Practice at Princeton University. He is the author of twenty books, including *This Endangered Planet, Security Through Disarmament,* and *The Vietnam War and International Law.* He has been a consul tant to the U.S. Arms Control and Disarmament Agency, the U.S. Senate Foreign Relation Committee, as well as a fellow of the Ford Foundation and the Center for Advanced Stud in Behavioral Sciences. Dr. Falk argues that our nuclear weapons technology has alread proliferated beyond the Nuclear Club and that we can either proceed as usual and face pos sible nuclear disaster, or choose complete denuclearization.

The nuclear age has brought us some standard nightmares. So far none has achieve reality, yet there is an uncomfortable feeling abroad that we are running out of time

*I wish to acknowledge the helpful suggestions of the editors and of my research associate, Jack Sande son.

TABLE 7–1 THE NUCLEAR CLUB AND POTENTIAL MEMBERS

Countries that have the bomb	Countries believed capable of building a nuclear bomb	Countries that could have the bomb within six years	Countries that could have the bomb in seven to ten years
United States	Canada	Argentina	Egypt
Soviet Union	West Germany	Australia	Finland
United Kingdom	Israel	Austria	Libya
France	Italy	Belgium	Yugoslavia
China	Japan	Brazil	
India	Pakistan	Denmark	
	South Africa	Iraq	
	Sweden	South Korea	
	Switzerland	Netherlands	
		Norway	
		Spain	
		Taiwan	

that sometime in the years ahead, perhaps not very far ahead, we will experience a nuclear catastrophe. One version of this dismal prospect involves the spread of nuclear weapons beyond the seven or eight governments (United States, Soviet Union, United Kingdom, France, and China, with demonstrated capacity, plus Israel and India, and possibly South Africa) that now either possess them or could in a matter of months (see Table 7–1).

The Crazy Leader Nightmare. One form of the proliferation nightmare is to imagine an aggressive and irrational leader's acquiring nuclear weapons. In the West this nightmare has often been associated with Third World leaders, for instance, Idi Amin of Uganda or Qadaffi of Libya. Such a crazy leader is seen as being willing to light the nuclear fuse that will ignite a nuclear war.

The Terrorist Nightmare. A second nightmare has to do with a terrorist group's somehow acquiring nuclear weapons, and then using them either to blackmail or shock civil society into submission. A physics undergraduate student at Princeton demonstrated a few years ago that virtually anyone with some money and modest technical skills could fabricate a bomb. The implications are evident, especially in a world where nuclear mercenaries—engineers with requisite training and experience—are available for hire.

A Nuclear Crowd. A third major image is the rapid spread of weapons to countries that presently have nuclear technology and could have the capacity to produce bombs and the missiles and planes to deliver them in the next four years. Joseph Nye, chief architect of Carter's nonproliferation policy, estimates that forty countries now fall into this category, and as many as a hundred countries could have a nuclear weapons potential by the end of the century. If a few of these countries break nonproliferation ranks and acquire nuclear weapons in the next several years, there may well be no way to restrain a surge of proliferation. Every government will come to associate the possession of nuclear weapons with the full exercise of sovereign rights.

Many people believe that there is no way to get rid of nuclear weapons altogether, as existing nuclear powers are unwilling to enter into far-reaching disarmament schemes. The best that can be done is to stop the problem from getting worse by preventing the spread of weapons to crazy leaders, terrorists, and a surge of new governments. In essence, avoiding nuclear war rests on the dynamics of deterrence (an exchange of credible nuclear threats) at the global level and on the prevention of proliferation (and hence a denial of nuclear capabilities) at the local and regional level. Whether this is a secure peace system is of course a matter of intense controversy, but what seems beyond debate is that some such dual order is likely to persist for at least the next decade or so.

Increasingly in Europe and the Third World the premises of a nonproliferation policy are called into question. For instance, are nonproliferation goals more important than commercial opportunities or relations with friendly foreign governments, as in U.S. objections to West Germany's sale of reprocessing facilities to Brazil? Such questioning supposes that proliferation is inevitable or is impossible to impede, and that the effort to do so results in self-inflicted commercial wounds. On quite different grounds the prominent Kenyan social scientist Ali Mazrui calls on Third World countries to pursue nuclear weapons capabilities partly to shock nuclear weapons countries into taking disarmament more seriously and partly to overcome the humiliation of being second-class citizens in a world of sovereign states.

The status of commercial nuclear technology must also be considered. Is generating electricity from commercial nuclear power sufficiently beneficial to be worth its proliferating implications? Can its beneficial applications be separated from easy diversion for weapons purposes? Can we ever get a handle on nonproliferation without renouncing nuclear technology altogether? Or contrariwise, can we ever hope to achieve the goals of nonproliferation unless we provide *all* states with equal access to commercial nuclear technology? As matters have worked out, compromises have been developed over the years yet the questions have not disappeared.

The Quest to Halt the Spread of Nuclear Weapons

At the outset of the nuclear age the United States possessed a nuclear weapons monopoly. In 1946 the U.S. government proposed that international society, under the auspices of the United Nations, control this deadly technology. After this proposal, popularly known as the Baruch Plan, was rejected by the Soviet Union, the United States shifted its emphasis to the maintenance of its nuclear monopoly, refusing to share the technology even with our allies. The Atomic Energy Act of 1946 prohibited the export of technical information in the atomic area altogether. It was assumed from the beginning that information on peaceful applications of nuclear energy could be used to develop a weapons program.

When the Soviet Union exploded its first nuclear device in 1949 it became obvious that any industrial power that was determined to become a nuclear power could do so. American policy again shifted from protecting its monopoly to discouraging military applications by unfriendly or uncontrollable nations. The essence of the new outlook, which persists to this day, is that the quest of additional countries for nuclear weapons is *separable* from the development of peaceful programs which are beneficial. In a speech to the U.N. General Assembly in 1953 President Eisenhower unveiled the much-heralded Atoms For Peace program. It reversed earlier policy, and offered American technical assistance in the nuclear area, including, if necessary, nuclear fuel, provided the recipient would make adequate commitments not to divert such information to a weapons program.

In 1957 the International Atomic Energy Agency (IAEA) was established to supervise a safeguards program designed to assure that sharing atomic information would not result in proliferation. But as additional countries opted for nuclear weapons (United Kingdom, 1957; France, 1960; China, 1964), pressure mounted among most leading governments to do something more dramatic to inhibit proliferation. The Nonproliferation Treaty (NPT) of 1968 was the result.

The Nonproliferation Treaty The NPT represents a twofold bargain between the nuclear and nonnuclear powers that deserves careful scrutiny. First, the nuclear powers agree to facilitate peaceful programs in nonnuclear states in exchange for the acceptance of IAEA safeguards at all nuclear facilities and for the renunciation of a weapons option. Second, the nuclear weapons states commit themselves "to pursue negotiations . . . relating to cessation of the nuclear arms race at an early date and to nuclear disarmament." The NPT has been ratified by more than 100 governments. No country party to the treaty has, in a strict sense, violated its terms.

At the same time the NPT has not been altogether successful. For one thing, China, France, and many threshold countries (those contemplating the acquisition of nuclear weapons) refused to sign on, preserving their discretion. Included among the prominent nonsignatories are Argentina, Brazil, India, Israel, Pakistan, South Africa, and Spain, countries which now include those with a demonstrated (India) or strongly suspected (Israel, South Africa) weapons capability.

Second, there are some loopholes within the treaty itself. Safeguards are obligatory only if the nuclear technology is transferred from one country to another. Indigenous facilities are not covered, such as those India relied on to produce weapons-grade material used for its test explosion in 1974. Under some circumstances, also, the NPT is a double fraud. On one hand the NPT penalizes those countries which have signed by restricting their development of commercial nuclear power, thus putting them at a disadvantage to those which have not signed the treaty. On the other hand even after countries sign the NPT there is no guarantee that the proliferation of nuclear weaponry will be restricted.

Third, the very terms of the NPT contribute to proliferation. It is now evident that the nuclear powers' promised help with peaceful programs provides a technical foundation on which the nonnuclear countries could build weapons. Because of limited natural uranium supplies there is growing pressure to satisfy energy requirements through fast breeder technology. The breeder is attractive because it creates more fuel than it consumes, but it is objectionable from a proliferation standpoint because it produces large quantities of plutonium. The introduction of "a plutonium economy" with the breeder reactor is frequently regarded as being as dangerous for world order as nuclear weapons themselves. Therefore to the extent that the NPT accelerates the spread of plutonium by the development of the civilian breeder reactor, it provides the material basis for proliferation. In the background, although rarely admitted, is the tension implicit in U.S. policy between enlarging its share of the export market for nuclear technology and inhibiting proliferation. In fact European proponents of breeder and reprocessing technology often claim that the United States uses nonproliferation arguments as a smokescreen to protect its competitive dominance over the export market for light-water reactors.

Fourth, the central feature of the NPT bargain seems to have collapsed. Increasingly, nonnuclear countries, especially in the Third World, believe that the main nuclear powers have no intention of disarming, or even of ending the arms race going on between them, especially since the breakdown of the SALT negotiations in 1979. As a consequence of this failure there is almost no moral force underlying the nuclear powers' nonproliferation pledge. Any nonnuclear country now feels justified in acquiring nuclear weapons if such a capability will strengthen its security. Furthermore, the leadership of the nuclear powers in such matters, especially the United States, has suffered as a result of what is widely viewed in the Third World to have been an NPT hoax, part of a wider effort by the superpowers to keep non-Western countries from achieving their place in the geopolitical sun.

Fifth, the NPT lacks any enforcement capability. It is an entirely voluntary arrangement. Article X even lets countries back out, provided only that three months' notice is given and that a withdrawing government refers to "extraordinary events" that it believes to have "jeopardized its supreme interests."

Other Strategies   Proliferation can also be inhibited by geopolitical pressures brought on by a powerful state. During the Carter presidency, for instance, the United States seems to have dissuaded, at least temporarily, South Korea and Taiwan from acquiring reprocessing plants, which would have facilitated the conversion of spent fuel into weapons-grade material. Pakistan may also have been similarly dissuaded in this period, or at least slowed down. South Africa was encouraged to avoid an overt demonstration of a weapons capacity. However, there are limits to such pressures.

Despite strenuous efforts the United States could not prevent the completion of a major deal where West Germany negotiated the transfer of full fuel-cycle technology, including a reprocessing technology, to Brazil. India's explosion in 1974 demonstrated that a bomb could be legally built from weapons-grade material diverted from a research facility not subject to any international regulation.

Superpowers can also inhibit proliferation by weakening the incentives to acquire nuclear weapons by a mixture of carrots and sticks. The main carrots are security guarantees (e.g., a mutual defense agreement, stationing of ground forces) and conventional weapons. Such carrots weaken the claim of such beleaguered countries as Taiwan and South Korea that nuclear weapons are needed for defense purposes. Sticks, in effect, threaten countries with reduced economic assistance or the removal of bases or troops. Again, carrots and sticks operate only in marginal settings. A determined government that falls within the threshold category cannot be stopped; Israel's acquisition of a presumed weapons capability is illustrative.

Nonproliferation can also be encouraged and supported by various methods to limit or ban nuclear weapons in international conflict. Nuclear-free zones could be established in particular regions where parties agree not to develop, possess, or allow the deployment of nuclear weapons on their territory. Or nuclear powers could pledge never to use nuclear weapons against nonnuclear states, which would essentially reward states for remaining in the nonnuclear category. Various arms control measures could also diminish the role of nuclear weapons in international diplomacy. However, even if these courses were feasible—which they are not in view of revived Cold War tensions—they would be a mixed blessing. In the short term at least, a diminished reliance on nuclear weapons by the superpowers might stimulate beleaguered states to acquire a nuclear deterrent of their own. One can imagine South Korea, Taiwan, Indonesia, Pakistan, Saudi Arabia, and even Japan moving to acquire nuclear weapons in the event that the United States removed its nuclear umbrella.

The ''success'' of nonproliferation policies is attributable largely to U.S. leadership; the Soviet Union has also been effective in preventing proliferation within its sphere of influence, excepting China. This relatively positive assessment, however, must be set off against the mounting dissatisfaction with the superpowers' nonproliferation policies as hypocritical and one-sided, perpetuating an imperial order in world politics without reducing the main dangers of general war. Increasingly, Third World opinion, despite its many diversities, tends to be highly skeptical of any nonproliferation policies given the failure of nuclear powers to disarm or restrict nuclear weapons. Perhaps indicative is the 1980 Lisbon Declaration, a document issued by an international group of concerned scholars who have put forward proposals for global reform. The Lisbon Declaration indicts the nuclear powers for "their own shameful record" regarding militarism and proposes "a more viable and equitable order" to replace the discredited nonproliferation treaty: demilitarization, whereby all nuclear weapons are renounced.

Nonproliferation in the 1980s

Other failures of nonproliferation are less blatant, but perhaps more serious. Arm-twisting of the sort that the United States undertook during the Carter years was ill-advised, unduly complicating relationships with friends and allies (for instance, Brazil, Pakistan, and South Korea) and giving up lucrative business opportunities to French and German competitors shackled by far looser nonproliferation scruples. The more conservative foreign policy leadership of Ronald Reagan's presidency appears to attach less significance to nonproliferation as a goal of foreign policy. While it will continue to pay lip service to general goals, it is far less likely to pressure other governments or to put the American nuclear industry at a competitive disadvantage vis-à-vis European exporters.

In the background, also, are the two administrations' differing attitudes toward nuclear power. The Carter liberals were reluctant supporters of nuclear power, slowing down all technological developments in the nuclear industry, such as breeder reactors and reprocessing plants, that would adversely affect efforts to discourage proliferation. The Reagan conservatives are preoccupied with increasing economic growth while emphasizing the supply side of energy, and seem far less disposed to limit the evolution of nuclear power in light of nonproliferation consequences.

Whether or not this policy shift will have any dramatic effects is difficult to assess at this stage. In the years just ahead the NPT will probably remain intact, but the wider nonproliferation regime may suffer erosion through this changed emphasis in the American leadership. Weakening the technical and political obstacles to nuclear weapons acquisition will certainly give more countries the capability to produce nuclear weapons, which in turn will give a wider range of governments a clear weapons option during the next decade. Whether this weakening process quickens the *pace* of proliferation or contributes to weapons acquisition by terrorist groups could be the most significant indicator of whether this new relaxed attitude toward proliferation is a mistake.

Even the most dedicated efforts objectively to appraise the prospects for war in a world of many nuclear nations are quite flimsy because reasoning depends so largely on "what if" suppositions. In the end it is a matter of worldview, including, implicitly, attitudes toward human nature and political institutions. As such, the nonproliferation debate has some resemblance to the controversy over gun control, and is likely to meet a similar fate, especially during these Reagan years.

The failure of nonproliferation is inevitable; only the timing is in doubt. It does not really matter, in this regard, whether nonproliferation Carter-style or Reagan-style prevails. The old models are not working, which induces leaders to embark on dike-plugging strategies that are ill-suited to the fallibilities and turbulence of human nature and societal institutions. Inherent to the problem is the fact that commercial and military nuclear power are *not* separable. "The link is a close one," said former

Nuclear Regulatory Commissioner Victor Gilinsky, "[and] an inconvenient reality that frequently intrudes on those who would deny it."

Two roads stand before us and we must make a choice. We can opt for a world with commercial and military nuclear capacities, which would mean that we will live permanently with intense latent proliferation. Not only would the dangers of nuclear disruption multiply through the years, eventually calling for diversion-proof technology and elaborate surveillance and verification machinery everywhere, but the effort to prevent such disruption might lead to the further curtailment of our liberties and to a deepening crisis of democracy. Or we can choose a path of denuclearization which, as a minimum, requires that we add nuclear disarmament to the serious foreign policy agenda and implies that we eventually renounce civilian nuclear power altogether.

Whether such an arrangement can be achieved is uncertain at best, but it seems that the economic burdens and war risks of a permanent arms race may make leaders and their citizenry more receptive than ever before to the appeals of denuclearization. At the very least, especially given the moral and legal problems associated with reliance on weapons of mass destruction for basic national security, it seems worth making an effort.

The Antinuclear Movement

DAVID DELLINGER

David Dellinger has been a pacifist and antiwar activist since World War II, when he resisted the draft and as a result served three years in prison. In 1968 he and other members of the Chicago 8 were put on trial for conspiracy charges in what became a landmark case epitomizing the deep divisions in our country concerning U.S. military involvement in Southeast Asia. Dellinger has taught at Yale University and is the author of several books, including *More Power Than We Know* and *Cuba, America's Lost Plantation*. He concedes that the antinuclear movement has a hidden agenda: that is, fighting for a society based on equality and democracy rather than dependence on energy monopolies and the nuclear industry.

In 1948 I took part in a sit-in in Washington, D.C., that urged the government to proceed with the commercial development of nuclear power. Like many people, I had been shaken by the message of the two tiny, still "primitive" atom bombs that had laid waste Hiroshima and Nagasaki. For the first time in the history of life on earth human beings now possessed the knowledge to destroy the human race and render the planet uninhabitable.

"Everything has changed now," Einstein reported, "except our thinking." Later he observed that "Each step appears as the inevitable consequence of the one before. At the end, looming ever clearer, lies general annihilation."

Now that the government was contemplating production of hydrogen bombs, a few dozen activists gathered in Washington to try to help the country change its thinking. We lobbied, held press briefings and public meetings, fasted for a week, and sat in at the Atomic Energy Commission's offices. We urged discontinuance of all development, testing, and deployment of nuclear weapons. We called instead for the peaceful uses of atomic fission through development of nuclear energy.

In succumbing to the lure of nuclear power we were influenced by our belief in the possibilities for human betterment through scientific discovery, and by our desire to see the burdens of poverty lifted from the destitute and hard-pressed, as the early advocates of nuclear power announced they would be. Also, rather than only saying *no* to something negative and diverse, we were eager to be able to say *yes* to something positive and uniting. We wanted to remind ourselves and our opponents that, as George Meredith once said, "The things that separate us from our fellows are as nothing compared to those that unite us with all humanity."

These were honorable intentions. They remain central to the practice and philosophy of the antinuclear, pro-renewable-energy movement today, though anyone whose information is limited to the mass media might not realize it. But we should have known better, even in 1948, than to believe what we were told about nuclear energy, because we knew that the government had lied to us about the necessity for dropping the bombs on Hiroshima and Nagasaki. To save thousands of *American* lives that would have been lost in an inch-by-inch, mile-by-mile invasion of Japan, it had said, we had to destroy Japanese lives. What the government failed to mention was that Japan had already sent out peace feelers, knowing that its military position was hopeless.

The government withheld this information from us because it was determined to demonstrate its terrifying new weapon before the war was allowed to end, the better to establish its intended domination of the postwar world. In all fairness, the men who were responsible for dropping the bomb did so in full confidence that U.S. hegemony is good for the world, even as it conveniently multiplies U.S. profits and privilege.

There were other reasons for a tiny governmental and corporate elite to make the momentous decision to lock the country into nuclear energy. Development of the Atoms for Peace program provided a front for the massive crash program in nuclear technology that was coveted by the Pentagon at a time when the public was both sobered by the "unthinkable" implications of nuclear warfare and insisting on postwar reductions in the military budget. For the utilities it provided billions of dollars worth of free research in governmental laboratories at taxpayers' expense.

As the country turned its face toward peace and looked for science and technology to usher in an era of beneficial abundance, government and industry continued the wartime practice of coopting some of the country's most inventive minds for a narrowly conceived program—heating water by a method that would have made Rube Goldberg blush—that simultaneously fed into the military and turned the dream of cheap, safe energy into a nightmare of bureaucratic bungling and coverups. Splitting the atom for "peaceful" purposes produced the plutonium for nuclear warheads and created the safety and security hazards that are beginning to make peace almost as dangerous as war.

We should have known that the information the military-industrial complex was circulating about the peaceful atom was distorted and incomplete. The government was acting on two dangerously undemocratic assumptions: that it knows what is best for the rest of us, including what we should be allowed to know; and that what is good for profits is automatically good for the country.

You may ask, What is the purpose of talking about the harmful effects of the drive for profits? Does it not confirm charges that the antinuclear movement serves as a front for people whose real goal is to do away with the things that made the country rich and strong: the profit motive and economic freedom? Does not the agitation and propaganda come mostly from malcontents who cannot make it in our society because they refuse to buckle down to work, hippies who do not believe in progress, and Communists who want to put the government in charge of our lives? Do not their very methods and attitudes give them away: sit-ins, plant occupations, lack of respect for "law and order" or private property?

Let me respond to these common accusations in two stages. First, I would like to talk about people in the antinuclear movement who initially favored nuclear power, as I did, but now oppose it. I will not list all the ways that producing nuclear energy is unavoidably fraught with peril, from mining uranium at the beginning of the cycle to storing wastes at the end with substances that are toxic for thousands of years. It is a cynical and frightening story. Those who tell it warn us from firsthand experience that the dangers are real, not invented or exaggerated by persons who hold a grudge against progress.

The information comes from people downwind and downstream of the plants— farmers, veterinarians, doctors, coroners, parents of children born dead or defective. It comes from medical personnel who have catalogued the incidents of cancer among the Indians who mine most of the uranium. Increasingly it also comes from scientists, engineers, and technicians who once advocated nuclear power.

Three top-level managers from General Electric's nuclear division sacrificed high-paying salaries when they resigned in 1976. Their reasons were clear when they testified before the Joint Committee on Atomic Energy of the U.S. Congress:

> We did so because we could no longer justify devoting our life energies to continual development and expansion of nuclear fission power—a system we believe to be so dangerous that it now threatens the existence of life on this planet. We could no longer rationalize away the fact that our daily labor would result in a radioactive legacy for our children and grandchildren for hundreds of thousands of years.

The range of people who speak from direct experience are today's counterparts of the Vietnam veterans and men and women still in the armed forces or government who played a vital role in educating the American people to the differences between the realities of U.S. policies in Vietnam and the pious claims of the government.

During the early years of the war antiwar activists were told that they were "stabbing the GIs in the back" by holding protest demonstrations, blocking entrances to the Pentagon, sitting in at draft boards and induction centers, and other activities similar to those for which the antinuke movement is frequently criticized today. It was, however, the generals and Washington politicians who resented these activities, not any significant number of GIs. As the GIs and other personnel returned from their tours of duty, more and more joined the protestors or organized dramatic actions of their own.

Today a similar confluence is taking place. The antinuclear movement flows out of the antiwar, civil rights, environmental, and social justice movements. It is renewed, as well as roiled up a bit, by young people who have not yet been sufficiently cowed to pretend that a naked emperor is fully clothed. It is constantly augmented and strengthened by experts and—sadly—victims, most of whom have no previous history as dissenters. Having experienced the evils of nuclear technology, they have been moved to speak and act in ways they once would have considered shocking. In recent years I have not taken part in a protest or sit-in that did not include someone from a family that had suffered stillbirths or leukemia, scientists such as those who resigned from General Electric, or plant workers with reports of coverups of frightening hazards. The movement has gained momentum and credibility by the support of these people who no longer believe the ruse of the peaceful atom.

Second, I would agree that many opponents of nuclear power, myself included, see the antinuclear effort as part of a larger struggle to change society. Our aim is to make the society more democratic. The path we advocate will help develop a closer, ecologically sound relationship with the natural universe of earth, air (wind), fire (sun) and water. We want the life-sustaining wealth that comes from nature, human labor, science and technology, to circulate throughout the entire population rather than being blocked as profits for a few multinational corporations.

Who among us would not prefer to play or listen to music, make love, read a book, or do any of a hundred things that life has to offer rather than attend a multitude of tedious political meetings? Who would not rather climb a mountain than a barbed-wire fence protecting a nuclear site? From the top of the mountain you look out on trees, lakes, rocks, and sky. From the top of the fence you see police waiting to pounce on you. You can expect to have your hands handcuffed painfully behind your back for hours, pay a fine or spend days, perhaps weeks, in overcrowded, unsanitary jails. All this and you can also offend some of your neighbors and be condemned as a misfit and troublemaker.

But to think that the oil cartel, the military, the banks, and the other corporate giants which back nuclear energy and arms will listen to sweet reason and the public interest without the help of an aggressive people's resistance movement is like thinking that the Polish workers would get satisfaction from the Polish Communist party by writing letters to the editor. Does anyone think that blacks in the South could have won even the right to vote or drink a cup of coffee in a "whites only" lunch counter without sit-ins, marches, and nonviolent civil disobedience?

Martin Luther King once told me that he never took part in an act of nonviolent resistance that did not initially offend those whom he was trying to enlist in the cause of racial justice, including many who claimed to believe in it but were doing nothing to advance it. Such acts, he believed, were necessary to raise the level of public debate and have an impact on public policy.

Similarly, Bertrand Russell, at the height of his fame and considered by many to be "the world's greatest living philospher," tried for years to alert the public to the dangers of atomic testing and the proliferation of nuclear weapons. So long as he limited his efforts to speeches and articles he made little progress. But when he was arrested for sitting down in London's Trafalgar Square with the Aldermaston Peace Marchers, his message was carried around the world and the antibomb movement took off.

The accident at Three Mile Island gained far more attention and was less successfully covered up than the worse accident at the Fermi plant near Detroit in 1966. The difference was that in the intervening years dramatic mass actions at Seabrook, New Hampshire, Diablo Canyon, California, Barnwell, South Carolina, and elsewhere had raised the level of public awareness to the point where it was harder to claim that everything was under control. Who would have learned that the cancer rate was higher downwind from the Rocky Flats, Colorado, nuclear weapons facility than upwind, if "troublemakers" had not made a public issue of the plant's existence?

One problem is that in matters of injustice and danger that we cannot see, we tend to ignore the implications of what we know. In our sophisticated state of alienation and passivity, primitive survival mechanisms often act to protect us from the unsettling knowledge of danger rather than from the danger itself.

After the fall of Hitler, Americans were shocked at the number of "good Germans" who said, with at least a grain of truth, that they had not fully understood what was going on in the neighboring death camps, despite the pervasiveness of the evidence. We want Americans to be fully aware, emotionally as well as intellectually, of the existence of factories of death in their neighborhoods that parade as sources of cheap, clean, safe energy. We want them to realize that continued reliance on nuclear "defense" means a high probability that computer error, national arrogance masquerading as patriotism, or simple political miscalculation in Washington or Moscow will start a war that would finish us all.

I write the foregoing with some hesitation. Passionate convictions about apocalyptic perils can lead well-intentioned individuals and movements to develop a self-righteous disregard for the rights, free will, and personal integrity of those who hold different opinions or no particular opinion at all on the issue. Such tendencies crept into the anti–Vietnam War movement toward the end of the 1960s.

Fortunately the lessons of the late 1960s and of the misunderstood explorations of the 1970s have led many activists to develop a closer relationship between the political and the practical, which has produced a movement that is deeply resistant to political zealotry and violence. More often than not activists go out of their way to communicate with those who are sympathetic but are skeptical about tactics of protest and resistance. Moreover, the growing personal involvement of important sectors of the antinuclear movement with the natural rhythms of alternate energy sources (woodstoves, windmills, solar, etc.) foster less frenetic, more respectful and harmonious relationships with the surrounding universe, both natural and human. Political demagogues and short-fused demonstrators encounter a calm resistance from a movement with these roots.

At its best the movement that advocates natural, renewable, safe sources of energy as a substitute for nuclear power is training people in the practice of direct or participatory democracy and in the spirit of respect for one's self, one's fellows, and one's natural environment, without which our society is doomed to continue its present downhill slide.

With apologies to T. S. Eliot, the world is in danger of ending with both a whimper and a bang. The whimpers we already hear are bound to increase, unless we learn to put common sense ahead of the seductive assurances of profit-seeking corporations, and begin to implement the slogan "People Before Profits" by cooperating to rise with our fellows instead of competing to rise above them. Increased development of nuclear power and weapons will lead to more cases of cancer, inflation, birth defects, unemployment, drug abuse, joyless work and joyless leisure, crime in the streets and suites, and ridiculous extremes of wealth and poverty.

"The world we have made," said Einstein, "as a result of the thinking we have done thus far creates problems that we cannot solve at the same level we created them at."

The antinuclear, pro-renewable-energy movement is flawed and uneven, as are all human endeavors. But it represents one effort to achieve new levels of thinking and living, and despite the reservations I expressed earlier, the effort is rewarding. If enough people join in such efforts, together we just might contribute to the survival of the human race and the earth we inhabit.

The Hidden Agenda

BERTRAM WOLFE

Bertram Wolfe has been vice-president and general manager of the nuclear energy division at General Electric Company since 1978. He is on the board of directors of the Atomic Industrial Forum (AIF) and the American Nuclear Energy Council (ANEC), and formerly on the board of directors of the American Nuclear Society (ANS). The author of several dozen popular and technical articles on energy, Dr. Wolfe has been active in various aspects of nuclear power research and management for the past twenty-five years. Dr. Wolfe argues that those who oppose nuclear energy are against *every* new energy source which is presently practical. Nuclear power, he maintains, far from being the issue at stake, is simply a vehicle by which activists hope to radically alter our political system.

Except for the creation of mankind, it is hard to identify a technical subject that has received more public attention and debate than energy. Indeed arguments about energy, and nuclear energy in particular, rival in intensity those about creation. I am not convinced that these public arguments illuminate the central issues. How can readers of this book, unfamiliar with the details of nuclear technology, evaluate the opposing views which are presented when supposed experts differ so radically on fundamental technical issues?

Consider briefly the subject of low-level radiation health effects. The air we breathe, the food we eat, and the ground we walk on are all naturally radioactive. Each of us is exposed by nature to a yearly background radiation dose of 100–150 millirems. Moving from a wood to a brick house would add about 15 millirems more and moving a hundred feet farther up a hill would add another millirem. The normal operation of a thousand nuclear plants and associated facilities would also add an additional millirem or so. It is known that an exposure a thousand times higher than background (100,000 millirems) produces detrimental health effects, but at natural background levels there is as much evidence of beneficial effects as there is of detrimental effects. Further, it is doubtful that we will ever know whether background radiation levels produce any significant health effects because the variable risks of normal living are so much greater.

And yet today not only the media but regulatory judges ponder this matter as though television interviews and learned legal arguments will reveal what nature refuses to disclose—even under subpoena.

The problem is that low-level radiation involves effects so small that its significance cannot be measured or determined. Thus public statements about low-level radiation represent judgments which can be colored, indeed determined, by philosophical considerations about the role of nuclear energy in society—considerations having nothing to do with the specific technical subject under discussion. Public discussions of low-level radiation effects, although presented as technical issues, frequently seem instead to be attempts to influence the public on societal issues which have nothing to do with radiation. How else does one explain the fruitless arguments as to whether an added yearly millirem of radiation exposure from long-lived reactor fission products will affect human beings 5000 years in the future; or alternatively, whether nuclear power will cleanse the earth of long-lived radioactive uranium decay products and save lives 100,000 years from now. More serious, how can one explain the purposeful generation of fear about the venting of gases from the Three Mile Island reactor building, when the resulting radiation exposure to nearby residents was equivalent to that received from a vacation in the mountains and the primary risk is from delays in cleanup of the stricken reactor?

The difficulty with much of the energy debate is that it focuses on technical issues such as radiation effects framed so that the central underlying philosophical questions are obscured.

Public concern about offshore oil leaks, the hazards of liquefied natural gas (LNG), the dangers of natural gas pipelines, western coal mining, nuclear waste disposal, environmental effects of shale oil, high-voltage transmission line effects, and the role of solar power, when considered in isolation, as is frequently the case, lead nowhere. The risks associated with each of these activities can be painted in colors of fear and emotion, but they can be meaningfully discussed only in terms of the alternative risks from alternative energy sources or from lack of energy. The question is not whether offshore oil drilling can produce oil leaks, but rather what means are to be used to meet the energy needs of the future and what kind of society will result.

One who believes that the future welfare of society is dependent on new domestic energy supplies will see large advantages to the development of nuclear power, offshore oil resources, and western coal, even at some risk and inconvenience. Those who believe that society suffers because we already use too much energy will not accept even minimal risk or inconvenience in order to supply more energy. A public discussion of energy development between groups with these opposing views is like a discussion of pork processing between farmers, meat processors, and Orthodox Jews and Muslims. One may talk about humane slaughtering techniques, but the underlying issue is whether or not pork should be eaten.

Contrary to most public perception, the basic disagreement over nuclear waste disposal is not over the risk, but rather over the benefits. A key benefit of a publicly acceptable waste disposal arrangement is the removal of an impediment to expanded use of nuclear power. Thus we find the nuclear advocates pressing for an early demonstration of a waste repository and the antinuclear forces arguing against even a demonstration with a few hundred spent fuel bundles in a military facility.

Most antinuclear groups are opposed to spent-fuel reprocessing, construction of spent-fuel storage facilities, and early waste repository construction. At the same time these groups argue that nuclear power should not be permitted unless means are available to accommodate the spent fuel. The issue may be couched in technical terms of "spent-fuel disposition," but in fact it is an argument over the morality of eating pork. The antinuclear groups ask not for safer nukes, not for more reliable nukes, not for more economical nukes—but only for *no nukes*.

But if there are to be no nukes, what will there be?

The major No Nukes organizations also oppose coal development, shale-oil development, LNG facilities, additional hydroelectric facilities, and offshore oil development. In the past they opposed exploitation of the present Alaskan oil fields and today they oppose exploration for new Alaskan oil.

When the California state hearings on the Sundesert nuclear plant were held, the Natural Resources Defense Council argued among other things that a coal plant should be considered instead of a nuclear plant. But after the Sundesert nuclear plant was cancelled and California Assemblyman Victor Calvo proposed legislation to ease coal plant siting, the NRDC wrote to Calvo. After describing a litany of concerns about coal, the NRDC concluded: "Coal is not a particularly desirable supply source from an environmental perspective, and we advocate that its use be minimized to the fullest extent possible."

Dr. Richard T. Kennedy, while an NRC commissioner, made the following observations: "Today a number of dedicated and sincere individuals believe that nuclear power should be abandoned. They are not just attempting to make nuclear power safer, not just seeking greater public oversight, not trying to keep nuclear plants at a distance. Instead, much of the opposition to nuclear power arises from what is a basic social view that the country's energy future should not be based on a technology as complex as nuclear power. The issue focuses more and more on the question of 'growth versus nongrowth.' I am not suggesting that to espouse and argue vigorously such views on this important matter of public policy is inappropriate. My problem is with the role that a regulatory body should play in response to this kind of contention."

There is no argument about the desirability of developing renewable solar resources. Almost everyone, including myself and my company, advocates their development. But as the reader can verify by getting an estimate from a local solar contractor, even the simplest solar technology—solar heating—is not yet here. Windmills are still losing their blades in high winds, and it is not clear whether large-scale biomass conversion is practical, or even a net energy producer.

The argument on solar goes much deeper: a closer look will indicate that those who advocate the near-term conversion to a solar energy economy coupled with the abandonment of presently available energy sources are in fact proposing to change society without explicitly indicating their intent.

Edward Nichols, associate editor of the *San Diego Union* newspaper, began his report of a major solar conference (over 1000 delegates) with these words:

It's possible to use sun power to heat your bath water without subscribing to the whole granola ethic, the *Washington Post* concluded recently.

It is, and many do. However, it also is true that most of those out in front of the organized movement for solar energy believe otherwise. They know that solar energy has a large role to play in America's energy future, but they also believe that they can substantially restructure American society through the medium of sunlight.

Their goals were starkly outlined recently at the second Annual Citizens' Solar Energy Conference at the University of Colorado at Boulder. As reflected in speeches and resolutions, the goal of the social reformers organized around the solar lobbies roughly is to:

○ Force utility companies to finance their own extinction.
○ Use an excess profits tax on big oil companies to finance their energy competitors so they, too, become dodo birds when the oil age is over.

○ Eliminate all nuclear power.
○ Reduce the production and distribution of energy to the lowest possible denominator—to local energy communes if possible.
○ Have "renewable resources" the main source of energy in the United States.

Clearly, solving the technical problem of producing economical solar energy is only a minor goal of much of the solar lobby.

It is not possible to characterize en masse the various "No Nukes," but there appear to be three major recurring themes in energy discussion.

The first is a general distrust of a society with abundant energy supplies. We find Stanford professor Paul Ehrlich, an antinuclear environmental spokesman, stating: "In fact, giving society cheap abundant energy at this point would be the equivalent of giving an idiot child a machine gun." Amory Lovins of Friends of the Earth puts it this way: "If you ask me, it'd be a little short of disastrous for us to discover a source of clean, cheap, abundant energy because of what we would do with it. We ought to be looking for energy sources that are adequate for our needs, but that won't give us the excesses of concentrated energy with which we could do mischief to the earth or to each other."

Consistent with this view is the position that any risk, inconvenience, or compromise is too high a price to pay for energy which in any event is apt to do harm. In February 1979, in a speech given in Charleston, South Carolina, the Interior Department's solicitor, Leo Krulitz, pleaded with environmental groups, "All we ask is your cooperation as we balance environmental concerns against the need to tap the vast potential of the U.S. Outer Continental Shelf to meet our critical energy problems." The response from attorney Bruce Terris, who frequently represents the No Nukes organizations, was that balance "involves compromising and sacrificing. That's their role. Our role is not to balance. Our role is to assert that the law be carried out."

What this means in fact is continued opposition and costly delays due to endless litigation. Krulitz indicates that this litigation is the "biggest threat to the administration's outer continental shelf program" for development of new oil and gas supplies. The opposition still continues.

A second theme is that society should be forced to alter and reorient itself to minimize energy use. Higher energy prices through resource severance taxes, onerous financial penalties to those deemed to use too much energy, the requirements that more expensive but more energy-efficient appliances be utilized, the elimination of free workplace parking, mandatory indoor summer and winter temperature limits, the control of household appliances from remote switching stations, a change by part of the population to nighttime living activities through imposed time-of-day utility rates, and the expanded use of manual labor are some of the vehicles proposed to help achieve this goal, in addition to constraint of supply through opposition to construction of new facilities.

A third theme is a general dissatisfaction with our present social and economic structure and the suggestion that energy should be used as a means for societal change not directly connected with energy.

Barry Commoner, at a major antinuclear rally in New York City on September 23, 1979, hardly spoke about nuclear power in his talk:

"Well, here we are—all 200,000 of us. Why? Who owns the air? Who owns the water? Who owns the earth? Who owns the sun? You know the right answer—we do, the people of the United States. But who controls our resources? The electric utilities, the oil companies—they decide whether we get radiation with our power. They decide whether we get carcinogens in our food. Who owns America? For whose benefit is this country run? For people or profits? Where can we find the political power? I'll tell you where. Right here—here are the antinuclear alliances, the Shads, the Clams, the Catfish, the Abalones that are forcing the nuclear industry to its knees. We stand for stopping nuclear power now! We stand for solar energy now! We stand for rolling back the price of fuel, the public control of the oil companies, for running this country in the interest of the citizens, not in the interest of profits. Now we can begin the battle to return to the citizens of this country the power that is rightfully theirs—to govern our own resources, our own lives, our own country. Thank you.

Barry Commoner proposes to move away from capitalism; Ralph Nader advocates a "consumer-controlled" economy; the Friends of the Earth argue for a steady-state economy of a form hardly recognizable from present-day America; and Jane Fonda and Tom Hayden tell us that "the stink in our midst is called Corporate Capitalism" and that the answer is a new economic system of public planning and public control called "Economic Democracy." The goal is to change society, although the changes sought are not always consistent among the various antinuclear forces.

As with the "No Nukes," it is impossible to categorize all the nuclear advocates under one banner. But in general those in favor of nuclear power are in favor of development of other available energy sources such as coal and offshore oil, and also in favor of the development of the sources for the future such as solar and fusion energy. The energy problem is described in terms of diminishing supplies of domestic oil and gas and consequent overdependence on imported supplies, whose long-term dependence is questionable. The proposed solution is not to force a change in society, but to minimize *forced* changes by providing alternate energy supplies.

Philosophically most nuclear advocates believe that abundant energy is a key element of a productive and stable society. They point to the close correlation between energy and employment and between energy and Gross National Product. And they point to the almost universal, worldwide correlation between per capita purchasing power and per capita energy use.

Although the increasing affluence of the United States has not been without its problems, the pro-energy advocates point out that beneficial societal effects have accompanied this affluence. Discriminatory actions against Jews, Orientals, and other minorities have greatly diminished. Blacks and women have started to emerge from economic serfdom.

Nuclear advocates believe that to accomplish goals such as further improving the living conditions of the disadvantaged and cleaning up the cities, additional energy supplies will be required. To Amory Lovins's question ''Would it be so terrible to live on half our per capita energy usage as we did in 1960?'' the nuclear advocates reply that in 1960 we had twice as many people living in poverty as we do today.

Fundamentally, pro-energy groups argue that as world petroleum supplies diminish, the expanded use of nuclear energy and other energy sources will help prevent forced changes in our society and will provide a means for worldwide improvement in living conditions. They note that with increasing affluence and accompanying energy consumption, birth rates voluntarily decline. Pro-energy groups argue that there is little hope of improving the lot of humanity without the energy supplies central to improved standards of living and thus believe it is appropriate that some risk and inconvenience be accepted to obtain these supplies.

The foregoing discussion is not intended to suggest that the energy dilemma is devoid of significant technical, economic, and environmental issues. It is misleading, for example, to gloss over difficulties in the areas of nuclear wastes, nuclear proliferation, reactor safety analysis, and reactor economics on the basis that nuclear power is needed, whatever its failings. But public discussions of such difficulties can also be misleading when they start from the philosophic base described, for instance, by Lovins and Ehrlich that nuclear power would still be unacceptable even if all of its technical, social, and economic problems were solved.

One must differentiate between the identification of a technical difficulty and the suggested conclusion which may result more from philosophical desires than from technical considerations. For example, the permanent disposal of high-level nuclear waste can by law only be done by the federal government. Does it follow that because the government has not yet built a nuclear waste repository that nuclear waste is unmanageable and that nuclear power should be abandoned? Or does it follow instead that the government program should be strengthened, and impediments removed, so as to speed up the construction of a waste repository? If nuclear power is abandoned in favor of, say, coal, will the wastes from coal present a lesser problem? And if it is concluded that coal is not satisfactory, or that coal cannot make up the deficit from the abandonment of nuclear power, will it be easier to deal with lack of energy than with nuclear wastes? As with most problems in life, one must deal with alternatives, and balance the risks and benefits of each.

Unfortunately such balanced considerations are rare. Instead our public regulatory proceedings and public discussions tend to focus without perspective on the detailed risks and inconveniences of each proposed energy project and energy source. But the alternative to our imperfect energy sources is not a perfect source; there is none available. If we continue to place impediments in the way of development of available energy sources, the alternative we will have chosen is a changed society, limited by energy supply constraints.

This type of society has been described longingly by Friends of the Earth as a society of "elegant frugality." But if it comes to pass, will we agree with this characterization or will we find it one of too much frugality and not enough elegance? And will we be happy with the way the frugality is shared? More important, if it comes to pass, will it be a path that was chosen with a clear understanding of *its* risks and inconveniences?

Epilogue

It is hard to believe that a $200-billion industry with the potential to help solve the nation's energy problems could be on the rocks—politically, technologically, and commercially. Certainly the controversy surrounding nuclear power is unique; nowhere can we find a historical precedent for a technology that held so much promise but that is now virtually stalemated.

Many of the beleaguered nuclear industry's problems are caused by external factors over which it has no control: high interest rates, slower growth in electricity demand, and an unpredictable political climate, to name a few. A utility director's reluctance to commit upward of $3 billion to the construction of a nuclear plant is understandable: the fourteen-year project will be subjected to the whims of perhaps three presidents, seven Congresses and state legislatures, and public utility commissions that often display a clear antinuclear bias.

But mistakes were also made along the road to commercial nuclear success which, combined with some of these external factors, have led to the current impasse. We need to examine these errors if we are to understand, and eventually resolve, the nuclear imbroglio. Learning from past experience is crucial if we wish to avoid repeating history as we pioneer other potentially controversial technologies. Whether one is pro- or antinuclear is irrelevant at this point: *both* sides would agree that the nuclear industry is beset with problems. A strong case can be made that the public has overreacted to the dangers of nuclear power. But our main concern is to determine the source of the problems. The dream of having 1000 reactors on line by the year 2000 has been forever laid to rest, and the time has come to ask ourselves: What went wrong?

At least one problem began at the atom's inception, when the technology was developed at a feverish pace in wartime secrecy. As the atom was transferred from the military to the commercial sector, much of the "closed door" attitude remained. In the 1950s and early 1960s the federal government and a small number of private companies pushed ahead furiously with reactor development, fully believing that they were establishing a source of electricity that—unquestionably—would be both economically and environmentally acceptable to the American people. David E. Lilienthal, first chairman of the Atomic Energy Commission, was later remarkably candid about the reasons for such high hopes: "The basic cause, I think, was a conviction, and one that I shared fully, and tried to inculcate in others, that somehow or other the discovery that had produced so terrible a weapon simply *had* to have an important peaceful use. . . .''

This conviction led various groups with either a military, industrial, or scientific stake in the atom's commercial success to believe that their work was linked to a vital national interest or at the very least, national prestige. Such thinking led to a situation where the term "national interest" could be used to promote almost anything pertaining to nuclear power. No one challenged, for example, the assumption that national interest would actually require the federal government to largely underwrite the costs of civilian nuclear development or to limit the utilities' liability to $560 million in the event of a nuclear accident.

The public, told little about the risks attending reactor development, was naïve about radiation hazards and potential proliferation problems. Today many people feel that they were deceived by the atomic establishment; for years they were told that a serious reactor accident was virtually impossible, and when one did occur there was outrage. Scientists from those early years argue that people refused to listen to those in the establishment who said an accident was possible, and that people simply did not read the information which was made available, albeit in forbidding technical form. Whatever the reason for people's naïveté, the results are obvious today: people accept airplane crashes and dam bursts as a fact of life, partly because these accidents do occur, and also because people were informed of the possibilities beforehand. But for years the public believed that a nuclear reactor could not "crash"; when one did so did faith in the atomic establishment.

Another problem which manifested itself over the years was the faulty premise on which the entire reactor program was built: it was thought that once the first prototype reactor was in operation the technology would be complete. The announcement in 1963 that the first nuclear plant would be built without direct government subsidy was cited as proof that the technology was mature, although at least one major problem remained: the lack of a suitable waste disposal technique. "The developers of nuclear power," noted Irvin C. Bupp in *Energy Future,* "had too narrow a view of their task. In many instances they did indeed behave as though they were building and selling a Boeing 747, and leaving the development and construction of airports radar, and pilot skills to some later date."

With regard to reactor development, few effective checks and balances were established in those early years; no means were available, for example, for the public to appraise a federally funded commitment to a technology whose expense, complexity, and potential for disaster was staggering. Government and industry officials publicly asserted that reactors were entirely safe, and their opinions—offered with little substantiation—grew into convictions that went unchallenged from within the establishment. The AEC, conceived in part as a watchdog, became more interested in promoting the cause, and offered little critical assessment of commercial nuclear development. Although the AEC was disbanded in 1974 for precisely this reason the nuclear advocates contend that the successor NRC has *over*compensated for the AEC's lack of regulatory zeal. Since the mid-1970s the NRC has regulated the industry rigorously—to death, some believe—by enforcing thousands of rules and regulations, some of which are frankly redundant or conflicting. Whether the public has been protected as a result of this plethora of regulations is questionable, but the effects on the nuclear industry have been obvious: every added constraint deals an increasingly intolerable economic blow to the utilities, which are faltering as a result.

In recent years it has been argued that many problems have been solved or ameliorated; the AEC has been abolished, federal regulations are slowly being streamlined, and problems with reactor safety, for example, have been addressed as a result of the ECCS hearings and other public testimonies. But as recently as 1979 the accident at Three Mile Island revealed deficiencies in evacuation planning, operator training, and control room design, which further undermined public confidence in the industry's ability to operate reactors safely. Nor was it particularly reassuring to learn that, due to a bureaucratic snarl, an engineer's report detailing instructions to prevent precisely the sequence of events that occurred was never circulated throughout the industry.

However, the accident *also* revealed that because of the engineered safety systems—the strength in depth—no one got hurt. A meltdown was averted and the industry's response to the event was a hopeful sign that changes would be made: control room designs were updated, emergency procedures were improved, and private watchdog institutes were established to supervise and analyze industry activity. In two decades of reactor operation, the utilities had compiled a remarkable safety record that was unparalleled in any other large technology: only one significant accident, and no immediate casualties.

Moreover the accident brought to many people's attention a problem that the nuclear advocates have noticed for years: sometimes the press reserves a glaring spotlight for nuclear power alone. Certainly, for example, the absence of a comprehensive evacuation policy at TMI was disturbing, and warranted being brought to national attention. But it should also be realized that evacuation plans are not required for chemical spills or toxic gas releases, which are very dangerous and far more probable than a nuclear accident requiring evacuation. Although that fact alone does not exempt the owners of the TMI plant from their negligence, it beckons us to consider nuclear technology in light of other large technologies. It may well be that the strict standards demanded of nuclear operations—not by the federal government, but by an industry that cannot afford another accident—serve to make nuclear power today the most consciously monitored technology of the modern age.

So what now? It is probably not beyond our means to develop nuclear power *and* ensure public safety. But changes in attitude must occur. One obvious lesson is that all involved must be more candid about the facts. The many low-level radiation studies, which often yield conflicting results, underline an important feature of the nuclear debate: enough uncertainties in the data exist that scientists cannot predict the precise long-term effects of radiation. Extrapolations from the Hiroshima-Nagasaki data and other high-level studies to determine low-dose effects are still hypothetical. Scientists will never gain the public's confidence unless they can say "we can only be reasonably sure" when these uncertainties are involved.

But other changes are in order if the nuclear issue is to be resolved. Both pro- and antinuclear forces need to work toward a forum by which they can discuss forth-rightly the problems attending nuclear power. At this point there is little communica-tion, let alone scientific exchange, between the opposing factions. In their attempts to outslander each other, both sides tend to lose sight of at least one reasonable goal: consolidating scientific expertise to ensure a safe, economical form of energy.

"To a large extent," noted David E. Lilienthal, "it [the citizen protest against atomic energy] has been a protest against the misuse of science." He is probably right. People are not necessarily opposed to the technology; they just want to know that nuclear power is safe, and that those who run the operations are trustworthy.

Public pressure is fast becoming one factor that may well determine the fate of commercial nuclear power. In hundreds of townships, counties, and cities, citizens are voting on whether to allow the construction or operation of nuclear facilities within their communities. In 1976 a referendum was held in California which, if approved, would have severely restricted the construction of new nuclear plants in the state. (The initiative was defeated by a 2–1 margin.) That same year, when ERDA announced that Michigan was being considered as a site for a federal waste facility, three Michigan counties voted by 8–1 margins to recommend that their counties not be used for a disposal site.

The debate over commercial nuclear power is as much a political issue as it is scien-tific or economic. If enough people find nuclear-generated electricity acceptable in all its aspects, and if citizens feel comfortable living near reactors, atomic energy could make a substantial contribution to our fuel reserves. Nuclear power conceiva-bly could even eliminate our energy problems.

The first step in resolving the nuclear controversy is to bridge the gap between what the experts know and what the public needs to know. When the facts are pre-sented accurately and comprehensively, people can judge for themselves the risks and benefits of nuclear power. This should have been done thirty-five years ago, but it is still not too late.

Notes

Chapter 2. Radiation

1. K. Z. Morgan, "The Need for Radiation Protection," *Radiologic Technology* 44, no. 6 (1973): 385; id., "Population Exposure to Diagnostic X-rays and Resultant Damage Can Be Reduced to 10% of Their Present Levels While at the Same Time Increasing the Quality and Amount of Diagnostic Information," testimony presented before hearings on H.R. 10790, Oct. 11, 1976; id., "Reducing Medical Exposure to Ionizing Radiation," Landauer Memorial Lecture, Stanford University, Sept. 27, 1974, in *American Industrial Hygiene Association Journal* 36 (May 1975): 358.

2. J. Rotblat, "The Risks for Radiation Workers," *Bulletin of Atomic Scientists,* Sept. 1978, p. 41, and references therein.

3. S. Jablon and H. Kato, "Childhood Cancer in Relation to Prenatal Exposure to Atomic Bomb Radiation," *Lancet* 14 (Nov. 14, 1970): 1000; B. MacMahon, "Prenatal X-ray Exposure and Childhood Cancer," *Journal of the National Cancer Institute* 28 (1962): 1173; R. M. Holford, "The Relation Between Juvenile Cancer and Obstetric Radiography," *Health Physics* 28 (Feb. 1975): 153; E. Landau, "Health Effects of Low-Dose Radiation: Problems of Assessment," *International Journal of Environmental Studies* 6 (1974): 51; R. H. Mole, "Ante-natal Irradiation and Childhood Cancer: Causation or Coincidence," *British Journal of Cancer* 30 (1974): 199.

4. I. D. J. Bross, "Leukemia from Low-Level Radiation," *New England Journal of Medicine* 287 (July 20, 1972): 107. See also, id., "Proceedings of a Congressional Seminar on Low Level Ionizing Radiation, House, 94th Congress" (published by the Environmental Policy Institute, Washington, DC 20003).

5. B. Modan et al., "Radiation-Induced Head and Neck Tumors," *Lancet* 1 (Feb. 23, 1974): 277.

6. T. F. Mancuso, A. Stewart, and G. Kneale, "Radiation Exposure of Hanford Workers Dying from Cancer and Other Causes," *Health Physics* 33, no. 5 (Nov. 5, 1977): 369.

7. S. Milham, *Occupational Mortality in Washington State, 1950–71,* 3 vols., NIOSH–76–175 (Washington, D.C.: U.S. Department of Health, Education and Welfare, 1974).

8. A. Stewart, speech at February 1978 meeting of the American Association for the Advancement of Science, Washington, D.C.

9. Depending on the specific form of cancer, either the linear, supralinear, or quadratic hypothesis could be appropriate in determining the radiation risk. There is mounting evidence in favor of a supralinear curve at low doses. But in some cases, and especially for experimental animal studies, the quadratic hypothesis may be more appropriate. For example, for breast cancer the linear hypothesis may exaggerate the effects of radiation at low doses. See K. Z. Morgan, "The Linear Hypothesis of Radiation Damage Appears to be Non-Conservative in Many Cases," *Proceedings, Fourth International Congress of the International Radiation Protection Association, Paris,* Aug. 1976, paper no. 451, p. 11.

10. See, for example, "NRC Covers Up Radiation Effects," by Ernest Sternglass published by Pacific News Service after the Three Mile Island incident, April-July 1979. Sternglass references me to assert that doses from strontium-90 and other fission products gave members of the public many times the government estimate. Sternglass never spoke to me before writing this article. He used a paper that I published in *Health Physics,* where I estimated the effects of a full fission product release: Less than one-millionth (!) of the iodines and other radiotoxic fission products escaped at TMI, as I verified for myself. Despite my letter of refutation of May 2, 1979, Pacific News Service still continued to publish the erroneous Sternglass article. See also my detailed calculations in

Health Physics, December 1980. See also R. E. Lapp, *The Radiation Controversy* (Greenwich, Conn.: Reddy Communications, Inc., 1979).

11. UNSCEAR, *Sources and Effects of Ionizing Radiation,* Report to the General Assembly, with Annexes (New York: United Nations, 1977); Committee on the Biological Effects of Ionizing Radiation (BEIR), "The Effects on Populations of Exposure to Low Levels of Ionizing Radiation," National Academy of Sciences, Washington, D.C., November 1972 (see also the 1980 report; hereafter cited as BEIR-I and BEIR-III); UNSCEAR, *General Assembly Official Records,* Twenty-fourth Session, Supplement no. 13 (A/7613) (New York: United Nations, 1969).

12. Ibid. See also B. Cohen and I. Lee, "A Catalogue of Risks," *Health Physics* 36 (1979): 707–22; R. L. Gotchy, "Estimation of Life Shortening Resulting from Radiogenic Cancer per Rem of Absorbed Dose," *Health Physics* 35 (1978): 563–65; C. A. Kelsey, "Comparison of Relative Risk from Radiation Exposure and Other Common Hazards," *Health Physics* 35 (1978): 428–30.

13. L. S. Taylor, *Radiation Protection Standards* (Boca Raton, Fla.: CRC Press, 1970). See also references in note 11, above.

14. A. Brodsky, R. E. Alexander, and R. J. Mattson, "NRC's Radiation Protection Regulations," in *Occupational Health Physics,* Proceedings of the Health Physics Society Midyear Topical Symposium, University of Colorado, Denver, 1976. See also Title 10, Chapter 1, *Code of Federal Regulations,* Parts 19, 20, 34.

15. J. Shapiro, *Radiation Protection* (Cambridge, Mass.: Harvard University Press, 1972).

16. B. L. Cohen, *Nuclear Science and Society* (New York: Anchor Books, 1974). See also id., "Impact of the Nuclear Energy Industry on Human Health and Safety." *American Scientist* 64 (1976): 550.

17. T. Straume and R. L. Dobson, "Leukemia and Cancer Risk Estimates from Recalculated Hiroshima and Nagasaki Doses," Lawrence Livermore National Laboratory, University of California, Livermore, Calif., paper presented at the 26th Annual Meeting of the Health Physics Society, Louisville, Ky., June 25, 1981. See W. E. Loewe and E. Mendelsohn, "Revised Dose Estimates at Hiroshima and Nagasaki," *Health Physics* 41, no. 4 (1981): 663–65; T.

Straume and R. L. Dobson, "Implications of New Hiroshima and Nagasaki Estimates: Cancer Risks and Neutron RBE," *Health Physics* 41, no. 4 (1981): 666–70.

18. See note 11, above.

19. Ibid.

20. Ibid.

21. "Population Dose and Health Impact of the Accident at the Three Mile Island Nuclear Station" (Washington, D.C.: U.S. Government Printing Office, May 10, 1979).

22. Cohen and Lee, "A Catalogue of Risks."

23. Cohen, *Nuclear Science and Society.*

24. T. F. Mancuso, A. M. Stewart, and G. W. Kneale, "Radiation Exposures of Hanford Workers Dying from Cancer and Other Causes," *Health Physics* 33 (1977): 369–85. See also G. W. Kneale, A. M. Stewart, and T. F. Mancuso, Letter to the Editor, *Health Physics* 36 (1979): 87.

25. T. W. Anderson, "Radiation Exposures of Hanford Workers—A Critique of the Mancuso, Stewart, and Kneale Report," *Health Physics* 35 (1979). See also K. J. Rothman, invited statement in "Public Meeting on the Low-Level Effects of Ionizing Radiation," sponsored by the U.S. NRC, Washington, D.C., April 7, 1978, pp. 80–83; B. S. Sanders, "Low Level Radiation and Cancer Deaths," *Health Physics* 34 (1978): 521–38; S. Marks, E. S. Gilbert, and B. D. Breitsenstein, "Cancer Mortality in Hanford Workers," paper presented at the *International Symposium on Late Biological Effects of Ionizing Radiation* (IAEA SM-224), Vienna, March 13–17, 1978; A. Brodsky, "A Statistical Method for Testing Epidemiological Results, as Applied to the Hanford Worker Population," *Health Physics* 36 (1979): 611–28; J. A. Reissland, "An Assessment of the Mancuso Study," National Radiological Protection Board, NRPB-R79, London, Sept. 1978.

26. Brodsky, "A Statistical Method for Testing Epidemiological Results."

27. As taken from my testimony before the House Subcommittee on Health and the Environment, Feb. 8–9, 1978.

28. E. S. Gilbert and S. Marks, "Cancer Mortality in Hanford Workers," *Radiation Research* 79 (1979): 122–48; G. B. Hutchison, B. MacMahon, S. Jablon, and C. E. Land, "Review of Report by Mancuso, Stewart and Kneale of Radiation Exposure of Hanford Workers," *Health Physics* 37 (1979): 207–20.

29. Dr. Barkev S. Sanders has also refuted Mancuso's claims that the AEC discounted the study because of "positive findings."

30. The adequacy of present safety standards has been further supported in detail in B. L. Cohen's article, "Impact of the Nuclear Energy Industry on Human Health and Safety."

31. Lapp, *The Radiation Controversy*. See also P. Beckman, *The Radiation Bogey* (Boulder, Colo.: Golem Press, 1980); id., *The Health Hazards of Not Going Nuclear* (Boulder, Colo.: Golem Press, 1978); Sir Fred Hoyle, *Energy or Extinction* (New York: Heinemann Educational Books, 1977).

32. J. W. Gofman and A. R. Tamplin, "Low Dose Radiation and Cancer," paper presented at the 1969 Institute for Electrical and Electronic Engineers Nuclear Science Symposium, San Francisco, Oct. 29–31, 1969 (*IEEE Transactions on Nuclear Science,* part I, vol. NS-17, pp. 1–9, Feb. 1970).

33. BEIR-I, The Committee on the Biological Effects of Ionizing Radiation (1972), "The Effects on Populations of Exposure to Low Levels of Ionizing Radiation."

34. G. W. Dolphin and I. S. Eve, "Some Aspects of the Radiobiological Protection and Dosimetry of the Gastrointestinal Tract" pp. 465–74 in *Gastrointestinal Radiation Injury,* edited by M. F. Sullivan et al. (Amsterdam: Excerpta Medica Foundation, 1971).

35. BEIR-I (1972).

36. BEIR-III Draft Report (1979), p. 227.

37. UNSCEAR, *Sources and Effects of Ionizing Radiation.*

38. J. W. Gofman, *Radiation and Human Health* (San Francisco: Sierra Club Books, 1981).

39. (1) Gilbert W. Beebe, Hiroo Kato, and Charles E. Land at the Radiation Effects Research Foundation in Japan of the Hiroshima-Nagasaki survivors ("Studies of the Mortality of A-bomb Survivors: 6. Mortality and Radiation Dose, 1950–74," *Radiation Research* 75 [1978]: 138–201); (2) The data of Louis H. Hempelmann and co-workers at the University of Rochester on infants irradiated at essentially zero years of age for the enlargement of the thymus gland and followed to ages thirty to forty years (L. H. Hempelmann, W. J. Hall, M. Phillips, R. A. Cooper, and W. R. Ames, "Neoplasms in Persons Treated with X-rays in Infancy: Fourth Survey in 20 Years," *Journal of the National Cancer Institute* 22 [1975]: 519–30); (3) The data of Roy E. Shore at the New York University Medical Center on women irradiated for postpartum mastitis and followed for periods up to thirty-four years (R. E. Shore, L. H. Hempelmann, E. Kowaluk, P. S. Mansur, B. C. Pasternack, R. E. Albert, and G. E. Haugie, "Breast Neoplasms in Women Treated with X-rays for Acute Postpartum Mastitis," *Journal of the National Cancer Institute* 59 [1977]: 813–22); (4) The data of John D. Boice at HEW and Richard R. Monson of the Harvard School of Public Health on women who developed breast cancer as a result of fluoroscopic radiation in the course of pneumothorax treatment of tuberculosis, and followed for periods up to forty-five years beyond irradiation (J. D. Boice, Jr., and R. R. Monson, "Breast Cancer in Women After Repeated Fluoroscopic Examination of the Chest," *Journal of the National Cancer Institute* 59 [1977]: 823–32); (5) The data of R. E. Shore and co-workers on children's irradiation for ringworm of the scalp and followed for periods as long as forty-five years (R. E. Shore et al., Data communicated to the BEIR-III Committee, in Draft Report [1979], p. 668).

40. J. W. Gofman and A. R. Tamplin, "Epidemiological Studies of Carcinogenesis by Ionizing Radiation," *Proceedings of the Sixth Berkeley Symposium on Mathematical Statistics and Probability* (Berkeley: University of California Press, 1971), pp. 235–77; J. W. Gofman, "Cancer Hazard from Low Dose Radiation," Docket no. RM–50–3, Statement before the Hearing Board of U.S. Nuclear Regulatory Commission, submitted by the Sierra Club, Buffalo, N.Y., 1977 (also appeared as CNR Report 1977–9, Committee for Nuclear Responsibility, San Francisco, Calif.); id., *Radiation and Human Health.*

41. Gofman, *Radiation and Human Health.*

42. Ibid., Chapter 7.

43. Gofman and Tamplin, "Epidemiological Studies. . . ."

44. We can illustrate the much greater sensitivity of those young at irradiation by applying practical use of the cancer-doses listed in Table 2–4. Let us compare the delivery of 1 rem of whole-body radiation to infants at 0 years of age and delivery of 1 rem of whole-body radiation to males at 35 years of age. We shall consider 100,000 persons irradiated for each group. From the definition of the cancer-dose, and assuming that the linear relationship holds between dose and cancer induction (which we

TABLE 2–4 HOW MUCH RADIATION WILL
CAUSE CANCER? THE CANCER-DOSE FOR
EXPOSURE TO WHOLE-BODY IONIZING
RADIATION FOR MALES AND FEMALES IN
THE U.S.

Age in Years at Irradiation	Cancer-Dose in Person-Rems (Males)	Cancer-Dose in Person-Rems (Females)
0	65	69
5	71	80
10	88	104[a]
15	178	217
20	200	249
25	201	252
30	234	285
35	328	399
40	538	636
45	1,200 (approx.)	1,400 (approx.)
50	13,000 (approx.)	15,000 (approx.)

Source: Abbreviated from Gofman; *Radiation and Human Health,* Chapter 8, where all the calculations which underlie the table are presented in detail.

[a] For example, if a population of women at the age of ten were irradiated with 104 person-rems, there will be one cancer death.

shall address below in showing that this by no means overestimates the effect), we calculate, for the irradiation of infants, as follows:

$$100{,}000 \text{ persons} \times 1 \text{ rem} = 100{,}000 \text{ person-rems, and}$$

$$\text{Cancer-dose (from Table 2–4)} = 65 \text{ person-rems per cancer death}$$

Therefore, lifetime yield of radiation-induced cancer deaths

$$= \frac{100{,}000 \text{ person-rems}}{65 \text{ person-rems per cancer death}} = 1538 \text{ cancer deaths}$$

We calculate, in a similar fashion, for those 35 years of age at irradiation, 100,000 person-rems received and a cancer-dose (from Table 2–4) of 328 person-rems per cancer death, the lifetime yield of radiation-induced cancer deaths

$$= \frac{100{,}000 \text{ person-rems}}{328 \text{ person-rems per cancer death}} = 305 \text{ cancer deaths}$$

Thus the lifetime cost in extra cancer deaths is 1538 ÷ 305, or 5.04 times as high for irradiation in infancy as at 35 years of age.

If we now examine the BEIR-III Final Report we find this remarkable conclusion:

There is now considerable evidence in near all the adult human populations studied that persons irradiated at higher ages have in general a greater excess risk of cancer than those irradiated at lower ages, or at least they develop cancer sooner [p. 38].

The added caveat in the BEIR-III statement, that "at least they develop cancer sooner," serves only to obscure the reality of the greate number of cancers for the lifetime of the exposed young population sample compared with that for the exposed older population.

To be sure, in a limited follow-up period of, say, fifteen years, one may well find fewe radiation-induced cancers in those who were young at irradiation, simply because both spo taneous and radiation-induced cancers have a lower incidence at young ages than at older ages. But if the lifespan is considered, many more radiation-induced cancers will develop among those irradiated while young, and it would be a cruel hoax to misuse the short-terr observations to obscure the much more seriou fate of those irradiated at an early age.

45. I thought I had seen everything in the arma-
mentarium of those engaged in wishing away the serious effect of low-dose radiation *until* the appearance of BEIR-III (both Draft and Final Reports). In the BEIR-III Draft Report, the following all-time amazing statement is to be found: "For ages under 10 years at expo-sure, the relative risk ratios obtained appear unreliable, and the ratios for ages 20–29 year at exposure are substituted for them" (p. 324

In a lifetime of scientific research I have seen many illustrations of the mishandling of scientific data, but this is by far the worst. In effect the BEIR Committee is saying: "We d not like the large percent excess cancer rate p rem in the zero- to nine-year age group becau it does not have the statistical base we would like. Instead, we will find a group, *grossly le sensitive according to every item of evidence available,* and substitute the value from that group for the high value that makes us uncom fortable. Please overlook the fact that, while we claim to be unbiased and authoritative, w have simply substituted one unreliable (becau totally inapplicable) value for the unreliable value we did not like."

With this type of handling of data one ca "prove" just about anything one would like prove. The committee members apparently realized what they had done, because they changed the Final Report *just a little.* "For ages under 10 years at exposure, the relative-risk ratios thus obtained appeared unreliable, and the ratios for ages 10–19 were substituted for them" (p. 247). The BEIR Committee is replacing one piece of scientific nonsense with one only slightly less nonsensical.

On the serious subject of breast cancer induction in those zero to nine years of age at irradiation, the BEIR Committee uses the very same study as a reference source to back up two totally contradictory statements. There were five cases of breast cancer (incidence, not deaths) in those zero to nine years of age at irradiation. Elsewhere I found that an analysis of these data suggested an extremely high sensitivity of the breast in this very young group to radiation-induction of cancer, although the numbers are still too small for statistical significance by themselves. The BEIR Committee apparently found these data extremely troublesome, and ended up by both acknowledging and denying the existence of these breast cancer cases.

On page 225 of the BEIR-III Final Report, it is stated: "Especially important are the absence of breast cancer among those who were under 10 in 1945, regardless of dose, and the absence of any excess breast cancer among those aged 40–49 at that time."

On page 337 of the same BEIR-III report, we find in a paper by Dr. Masayo Tokunaga of the Radiation Effects Research Foundation in Hiroshima and co-workers, based on the *same reference source,* the following statement: "Substantial evidence from controlled studies of increased breast-cancer risk in women exposed to ionizing radiation before the age of ten is lacking. Only one (nonexposed) breast cancer was found in the 1950–69 LSS series among women 0–9 yr old at the time of bombing. Five cancers in the same age group were found in the 1950–1974 series, including one with a breast-tissue dose of 57 rads and four with doses of less than 20 rads" (M. Tokunaga, J. E. Norman, Jr., M. Asano, S. Tokuoka, J. Esaki, I. Nishimori, and Y. Tsuji, "Malignant Breast Tumors Among Atomic Bomb Survivors, Hiroshima and Nagasaki, 1950–1974," *Journal of the National Cancer Institute* 62 [1979]: 1347–59).

It is scientifically simply incredible that the very same reference by Tokunaga is cited to show *five* breast cancers (on page 337) as was

cited to show *zero* breast cancers (on page 225). Shades of the Great Houdini!

46. Gofman, *Radiation and Human Health.*
47. The governmental and nongovernmental sources estimated that the radionuclides released from the plant resulted in an aggregate population dose of between 5000 and 50,000 person-rems. If these numbers were to be translated into some sizable number of cancer fatalities, the industry realized it would be facing a very serious public relations problem. But by dint of a carefully prepared disinformation program, the requisite answers for Three Mile Island were available. UNSCEAR uses the absolute risk method, even though that method is thoroughly discredited by the evidence itself, and UNSCEAR estimates that for every 10,000 person-rems, there would be one fatal cancer. UNSCEAR's estimate, coupled with the government estimate that the radiation dose was 5000 person-rems (or lower), explains the widely heralded estimate that possibly there would be one-half of a cancer caused by the Three Mile Island accident (5000/10,000 is ½).
48. Gofman, *Radiation and Human Health.*
49. L. H. Hempelmann, W. H. Langham, C. R. Richmond, and G. L. Voelz, "Manhattan Project Plutonium Workers: A Twenty-Seven Year Follow-up Study of Selected Cases," *Health Physics* 25 (1973): 461–79; G. L. Voelz, L. H. Hempelmann, J. N. P. Lawrence, and W. D. Moss, "A 32-Year Medical Follow-up of Manhattan Project Plutonium Workers," *Health Physics* 37 (1979): 445–85.
50. G. L. Voelz, "What We Have Learned About Plutonium from Human Data," *Health Physics* 29 (1975): 551–61.
51. J. W. Gofman, "The Cancer Hazard from Inhaled Plutonium," *Congressional Record* 121 (1975): S 14610–16 (also as CNR Report 1975–1R); id., "The Estimated Production of Human Lung Cancers by Plutonium from Worldwide Fallout," *Congressional Record* 121 (1975): S 14616–19 (also as CNR Report 1975–2); id., "Testimony for the GESMO Hearings: General Response to All Critiques of John Gofman's Estimates of the Lung Cancer Hazard of Plutonium," submitted by the Public Interest Research Group, Washington, D.C. (also as CNR Report 1976).
52. ERDA Laboratory Critiques of Gofman's Plutonium Papers: (1) C. R. Richmond, "Review of John W. Gofman's Reports on Health Hazards from Inhaled Plutonium,"

ORNL/TM–5727, 1976; (2) D. Grahn, "Comments Prepared by Dr. D. Grahn," Division of Biological and Medical Research, Argonne National Laboratory, Argonne, Ill., 1975; (3) W. J. Bair, "Review of Reports by J. W. Gofman on Inhaled Plutonium," BNWL–2067/UC–41, Battelle Northwest Laboratories; (4) M. B. Snipes, A. L. Brooks, R. G. Cuddihy, and R. O. McClellan, "Review of John Gofman's Papers on Lung Cancer Hazard from Inhaled Plutonium," LF–51/UC–48, Lovelace Foundation for Medical Education and Research, Albuquerque, N.M., 1975; (5) J. W. Healy, E. C. Anderson, J. F. McInroy, R. G. Thoman, and R. L. Thomas, "A Brief Review of the Plutonium Lung Cancer Estimates by John W. Gofman," LA–UR–75–1779, Los Alamos Scientific Laboratory, Los Alamos, N.M., 1975.

53. Gofman, "Cancer Hazard from Low Dose Radiation."

54. R. E. Albert, M. Lippmann, H. T. Peterson, Jr., J. Berger, K. Sanborn, and D. Bohning, "Bronchial Deposition and Clearance of Aerosols," *Archives of Internal Medicine* 131 (1973): 115–27.

55. Gofman, "Cancer Hazard from Low Dose Radiation." A second comment frequently made concerning the possibility of long-term retention of plutonium in the bronchi is that mucus secretion will be sufficient to clear the plutonium out even if the ciliary action is damaged in cigarette smokers. This is pure, unsupported speculation. When it was convenient to argue that ciliary action *plus* mucus secretion are the mechanisms for prevention of plutonium accumulation in the bronchi, both were used as the explanation. Once I pointed out the damage to the cilia, then, without any support whatever in scientific evidence, the nuclear promotional community began to argue that mucus secretion alone would suffice to clear the plutonium. Some evidence beyond speculation would be helpful.

But all this has recently become quite moot with the appearance of strong experimental evidence supporting my contention of prolonged retention of insoluble particulates in the lungs of cigarette smokers. Drs. David Cohen, Satoaki Arai, and Joseph F. Brain of Harvard University, using Fe_3O_4 (a magnetic dust) as the insoluble material, demonstrated conclusively that insoluble particulates are retained in cigarette smokers to an extent five times greater

than in nonsmokers, even after one year beyond the inhalation ("Smoking Impairs Long-Term Dust Clearance from the Lung," *Science* 204 [1979]: 514–17). They did this by using a sensitive magnetic detector to measure the amount of magnetic dust left in the lungs of smokers and nonsmokers. While it is true that these new studies do not prove the retention is in the bronchi, it is reasonable to expect that at least part of the retention must be, since it is in the bronchi that one finds such major differences between smokers and nonsmokers. The new work of Cohen and co-workers represents a great embarrassment for those who argue against long-term retention of insoluble particulates in smokers.

Another argument used by the nuclear promoters is that I may not have estimated the surface area of the bronchi correctly in assessing the dose to be received from plutonium. It must be pointed out that we simply do not know accurately the exact part of the bronchial tree that is vulnerable to lung cancer development. It may well be that my estimate is too low or too high, and only much more data can ever answer this. One cannot exonerate plutonium with speculation about the true area of bronchial surface involved.

The most interesting aspect of all this is the ease with which a massive problem for the nuclear industry was swept under the rug with the one unsupportable assumption that no plutonium would be retained in the bronchi for the long term. If scientists learn how to make the "right" assumptions, virtually any sticky problem can be solved—by decree.

Last, in some of the ERDA Laboratory critiques of my work on plutonium it was suggested that the "hot particle theory" has been discredited. This criticism of my work is so ludicrous as to be incredible since at the outset of my papers on plutonium carcinogenicity I explicitly pointed out that *I am not using the "hot particle theory" in any way, shape, or form*. The "hot particle" concept of Dr. Donald Geesaman suggested that local high doses of radiation might be more effective *per rem* in producing cancer than low doses. My work makes absolutely no suggestion at all that the carcinogenicity per rem is increased by local high doses, so this is the most foolish of the red herrings used to attempt to criticize my calculations on plutonium carcinogenicity.

56. Natural radiation is 0.1 rem/year. This is (0.1

rem/year) × (0.01 joule/kg-rem) × (75 kg/person) × ($6 × 10^{12}$ Mev/joule) × (1 gamma ray/Mev) × (1 year/$3 × 10^7$ sec) = 15,000 gamma ray/sec.

57. According to the 1980 BEIR Report the cancer risk is $120 × 10^{-6}$/rem. This is ($120 × 10^{-6}$/rem) × (0.1 rem/year) × (1 year/$3 × 10^7$ sec) × (1 sec/15,000 gamma rays) = $1/(30 × 10^{15}$ gamma rays) = 1/(30 quadrillion gamma rays).

58. A. C. Upton et al., "Late Effects of Fast Neutrons and Gamma Rays in Mice as Influenced by the Dose Rate of Irradiation: Induction of Neoplasia," *Radiation Research* 41 (1970): 467.

59. D. E. Lea, *Actions of Radiation on Living Cells* (London: Cambridge University Press, 1955).

60. B. W. Fox and L. G. Lajtha, "Radiation Damage and Repair Phenomena," *British Medical Bulletin* 29 (1973): 16; M. M. Elkind and J. L. Redpath, chapter 3 in *Cancer,* edited by S. S. Bicker (New York: Plenum Press, 1975), p. 51. See also UNSCEAR, *Ionizing Radiation: Levels and Effects* (New York: United Nations, 1972); R. A. McGrath and R. W. Williams, "Reconstruction *in vivo* of Irradiated *Escherichia coli* Deoxyribonucleic Acid; the Rejoining of Broken Pieces," *Nature* 212 (1966): 534; C. D. Town, K. C. Smith, and H. S. Kaplan, "Influence of Ultrafast Repair Processes (Independent of DNA Polymerase I) on the Yield of DNA Single-Strand Breaks in *Escherichia coli* K-12 X-Irradiated in the Presence or Absence of Oxygen," *Radiation Research* 52 (1973): 99.

61. B. L. Cohen, "The Cancer Risk from Low Level Radiation," *Health Physics* 39 (1980): 659.

62. R. H. Mole, "Effects of Dose Rate Protraction: A Symposium," *British Journal of Radiology* 32 (1959): 497.

63. Cohen, "The Cancer Risk from Low Level Radiation."

64. Ibid.; R. L. Ulbrich, M. C. Jernigan, C. E. Cosgrove, L. C. Sutterfield, N. D. Bowles, and J. B. Storer, "The Influence of Dose and Dose Rate on the Incidence of Neoplastic Disease in RFM Mice After Neutron Irradiation," *Radiation Research* 68 (1976): 115; M. P. Finkel and B. O. Biskis, *Progress in Experimental Tumor Research* 10 (1968): 72; J. Marshall, Argonne National Laboratory, private communication, 1978; M. M. Elkind and G. F. Whit-

more, *The Radiobiology of Cultured Mammalian Cells* (New York: Gordon and Breech, 1967), p. 219.

65. B. L. Cohen and I. S. Lee, "A Catalog of Risks," *Health Physics* 36 (1979): 707.

66. Both estimates are analyzed on p. 539 of the 1977 UNSCEAR Report.

67. L. Ehrenberg, G. von Ehrenstein, and A. Hedgran, "Gonad Temperature and Spontaneous Mutation Rate in Man," *Nature,* Dec. 21, 1957, p. 1433.

68. J. V. Neel, J. Kato, and W. J. Shull, "Mortality of Children of Atomic Bomb Survivors and Controls," *Genetics* 76 (1974): 311.

69. J. Aurier et al., "Report of the Task Group on Health Physics and Dosimetry to the President's Commission on the Accident at Three Mile Island," Oct. 31, 1979; Mitchell Rogovin, "Three Mile Island," vol. 2, part 2, "A Report to the NRC Commissioners and the Public," p. 399.

70. B. L. Cohen and H. N. Jow, "A Genetic Hazard Evaluation of Low Level Waste Burial Grounds," *Nuclear Technology* 41 (1978): 381.

71. B. L. Cohen, "Hazards from Plutonium Toxicity," *Health Physics* 32 (1977): 359.

72. Thomas Dunkel, U.S. Dept. of Energy, private communication to B. L. Cohen, May 24, 1979.

73. National Safety Council, "Accident Facts," published annually.

74. Ibid.

75. Ibid.

76. The conclusion that electricity derived from coal burning is more injurious to human health than electricity generated from nuclear energy is supported by many published studies whereas there have been no known studies which arrive at the contrary conclusion. It is noteworthy that this list includes a report from the Union of Concerned Scientists, a strongly antinuclear group that serves as scientific advisor to Ralph Nader. Some of these studies are: (1) H. Inhaber, "Risk of Energy Production," Atomic Energy Control Board Report AECB–1119 (Ottawa), 1978; (2) R. L. Gotchy, "Health Effects Attributable to Coal and Nuclear Fuel Cycle Alternatives," NUREG–0332, 1977; (3) C. L. Comar and L. A. Sagan, "Health Effects of Energy Production and Conversion," in *Annual Review of Energy,* edited by J. M. Hollander, 1 (1976): 581; (4) L. B. Lave and L. C. Freeburg, "Health Effects of Electricity Generation from Coal,

Oil, and Nuclear Fuel," *Nuclear Safety* 14, No. 5 (1973): 409; (5) L. D. Hamilton, "The Health and Environmental Effects of Electricity Generation," Brookhaven National Laboratory, 1974; also "Energy and Health," in *Proceedings of the Connecticut Conference on Energy,* Dec. 1975; (6) B. L. Cohen, "Impacts of the Nuclear Energy Industry on Human Health and Safety," *American Scientist* 64 (1976): 291; (7) D. J. Rose, P. W. Walsh, and L. L. Leskovjan, "Nuclear Power—Compared to What?" *American Scientist* 64 (1976): 291; (8) S. M. Barrager, B. R. Judd, and D. W. North, "The Economic and Social Costs of Coal and Nuclear Electric Generation," Stanford Research Institute Report, March 1976; (9) D. W. North and M. W. Werkhofer, "A Methodology for Analyzing Emission Control Strategies, *"Comput. Ops. Res.* 3 (1976): 187; (10) K. A. Hub and R. A. Schlenker, "Health Effects of Alternative Means of Electrical Generation," in *Population Dose Evaluation and Standards for Man and His Environment* (Vienna: IAEA, 1974); (11) "Comparative Risk—Cost-Benefit Study of Alternative Sources of Electrical Energy," WASH–1224, U.S. Atomic Energy Commission Dec. 1974; (12) Union of Concerned Scientists, "The Risks of Nuclear Power Reactors," 1977; (13) Nuclear Energy Policy Study Group, "Nuclear Power—Issues and Choices" (Cambridge, Mass.: Ballinger, 1977); (14) R. Wilson and W. J. Jones, *Energy, Ecology, and the Environment* (New York: Academic Press, 1974); (15) National Academy of Sciences, Committee on Nuclear and Alternative Energy Systems, *Energy in Transition, 1985–2010* (San Francisco: W. H. Freeman, 1980); (16) American Medical Association, Council on Scientific Affairs, "Health Evaluation of Energy Generating Sources," *Journal of the American Medical Association* 240 (1978): 2193; (17) United Kingdom Health and Safety Executive, *Comparative Risks of Electricity Production Systems* (London: HMSO 1980); (18) Norwegian Nuclear Power System, *Nuclear Power and Safety* (Oslo: Universitetsforlaget, 1978); (19) W. Ramsay, "Unpaid Costs of Electrical Energy," in *Resources for the Future* (Baltimore: Johns Hopkins University Press, 1979).

77. U.S. Senate Committee on Public Works, "Air Quality and Stationary Source Emission Control," a report prepared by the National Academy of Sciences, 1975; see pp. 626, 631.

Many similar or higher estimates have been published, for example, L. B. Lave and E. P. Seskin, "An Analysis of the Association Between U.S. Mortality and Air Pollution," *Journal of the American Statistical Association* 68 (1977): 284; M. G. Morgan et al., *Energy Systems and Policy* 2 (1978): 287; S. J. Finch et al., in *Advances in Environmental Science and Engineering;* W. Winkelstein et al., *Archives of Environmental Health* 14 (1967): 162.

78. U.S. Nuclear Regulatory Commission, "Reactor Safety Study," WASH–1400, 1975.

79. Ibid.

80. Union of Concerned Scientists, "The Risks of Nuclear Power Reactors."

81. J. L. Lyon, M. R. Klauber, J. W. Gardner, and K. S. Udall, "Childhood Leukemias Associated with Fallout from Nuclear Testing," *New England Journal of Medicine* 300 (1979): 397.

82. J. W. Gofman and A. R. Tamplin, *Poisoned Power* (Emmaus, Penna.: Rodale Press, 1971).

83. "Radio-ecological Assessment of the Wyhl Nuclear Power Plant," Department of Environmental Protection of the University of Heidelberg, 1978–1979.

84. Natural Resources Defense Council, "Petition to Amend Radiation Protection Standards as They Apply to Hot Particles," submitted to U.S. AEC and U.S. Environmental Protection Agency, Feb. 14, 1974.

85. The Medical Research Council, *The Toxicity of Plutonium* (London: HMSO 1975); NCRP, "Alpha Emitting Particles in the Lungs," NCRP Report N. 46, 1976; National Radiological Protection Board, United Kingdom, Report R–29 and Bulletin N. 8, 1974; W. J. Beir, C. R. Richmond, and B. W. Wachholz, "A Radiological Assessment of the Spatial Distribution of Radiation Dose from Inhaled Plutonium," U.S. AEC, WASH–1320; National Academy of Sciences BEIR Committee, "Health Effects of Alpha-Emitting Particles in the Respiratory Tract," EPA 520/4–76–031, 1976.

86. Cohen, "Hazards from Plutonium Toxicity."

87. Based on 10,000 population per square kilometer; total size does not matter much. This is the population density within city limits of most large cities.

88. R. Nader, "Are Nuclear Side Effects Hazardous to Your Health," *Family Health,* Jan. 1977, p. 53.

89. J. W. Gofman, "The Cancer Hazard from Inhaled Plutonium," Committee for Nuclear Responsibility Report CNR 1975–1, 1975.

90. The paper was never published in a scientific journal, but it drew at least six scientific critiques (below) and was ignored by the ICRP, UNSCEAR, BEIR, NCRP, MRC, and NRPB, all of which have published reports on plutonium toxicity. I have never encountered a scientist active in work on radiation effects who supports the theory. See (1) Cohen, "Hazards from Plutonium Toxicity"; (2) W. J. Bair, "Review of Reports by J. W. Gofman on Inhaled Plutonium," Battelle Northwest Laboratory Report BNWL–2067; (3) C. R. Richmond, "Review of John W. Gofman's Report on Health Hazards from Inhaled Plutonium," Oak Ridge National Laboratory Report ORNL–TM5257, 1975; (4) J. W. Healy et al., "A Brief Review of the Plutonium Lung Cancer Estimates by John W. Gofman," Los Alamos Scientific Laboratory Report LA–UR–75–1779, 1975; (5) M. B. Snipes et al., "Review of John Gofman's Papers on Lung Cancer Hazard from Inhaled Plutonium," Lovelace Foundation (Albuquerque, N.M.) Report LF–51/UC–48, 1975; (6) "Comments Prepared by D. Grahn," Argonne National Laboratory, 1975.

91. J. W. Gofman, "Estimated Production of Human Cancers by Plutonium from Worldwide Fallout," Committee for Nuclear Responsibility Report No. NCR 1975–2. Also several papers in popular magazines.

92. B. L. Cohen, "Plutonium Containment," *Health Physics* 40 (1981): 76.

Chapter 3. Reactor Safety

1. U.S. Atomic Energy Commission, "Theoretical Possibilities and Consequences of Major Accidents in Large Nuclear Power Reactors," U.S. AEC Report WASH–740, 1957. A brief summary, in perspective, is given in A. V. Nero and M. R. K. Farnaam, "A Review of Light-Water Reactor Safety Studies," Lawrence Berkeley Laboratory Report LBL–5286, Jan. 1977.

2. See Anthony V. Nero, Jr.'s *A Guidebook to Nuclear Reactors* (Berkeley: University of California Press, 1979) for a detailed discussion of reactor types, nuclear fuel cycles, and nuclear safety, with many illustrations.

3. U.S. Nuclear Regulatory Commission, "Reactor Safety Study: An Assessment of Accident Risks in U.S. Commercial Nuclear Power Plants," U.S. NRC Report WASH–1400 (or NUREG–75/014) in nine volumes, Oct. 1975. A brief, but relatively substantial, summary of WASH–1400 is contained in Nero, "A Review of Light-Water Reactor Safety Studies." Nero's report also summarizes a report on reactor safety by the American Physical Society.

4. See H. W. Lewis et al., "Report of the Nuclear Regulatory Commission Review Panel on Probabilistic Risk Analysis," U.S. NRC Report NUREG/CR–0400, Sept. 1978.

5. Report of the President's Commission on the Accident at Three Mile Island—The Need for Change: The Legacy of TMI (Washington, D.C.: U.S. Government Printing Office, Oct. 1979).

6. Z. Medvedev, *Nuclear Disaster in the Urals* (New York: W. W. Norton, 1980); *Nuclear Safety* magazine, Oak Ridge National Laboratory, March–April 1979; A. C. Chamberlain, *Royal Meteorology Society Quarterly* 85 (1959): 354.

7. J. Beyea and F. von Hippel, *Nuclear Reactor Accidents: The Value of Improved Containment*, Princeton University, PU/CEES #94, 1980; J. Beyea, *A Study of the Consequences of Hypothetical Reactor Accidents at Barsebeck*, Swedish Energy Commission, 1978.

8. Union of Concerned Scientists, *The Risks of Nuclear Power* (Cambridge, Mass., 1977).

Chapter 4. Nuclear Waste Disposal

1. In the space of one essay it is impossible to contribute a meaningful and informative overview of so controversial a subject as nuclear waste disposal. For the reader who would like to fill in the many aspects which I have to omit, I recommend a series of studies prepared by: (1) the U.S. Geologic Survey—J. D. Bredehoeft et al., "Geologic Disposal of High-Level Radioactive Wastes—Earth-Science Perspectives," Geological Survey Circular 779 (Reston, Va.: U.S. Geologic Survey, 1978); (2) the Environmental Protection Agency—Report of an Ad Hoc Panel of Earth Scientists, "State of Geological Knowledge Regarding Potential Transport of High-Level Radioactive Waste from Deep Continental Repositories" (Washington, D.C.: U.S. Environmental Pro-

tection Agency, Office of Radiation Programs, 1978); (3) the Office of Science and Technology—(a) "Report to the President by the Interagency Review Group on Nuclear Waste Managements" (Washington, D.C.: Executive Office of the President, Office of Science and Technology Policy, March 1979), TID–29442; (b) Interagency Review Group on Nuclear Waste Management, "Subgroup Report on Alternative Technology Strategies for the Isolation of Nuclear Waste" (Washington, D.C.: Executive Office of the President, Office of Science and Technology Policy, Oct. 19, 1978). TID–28818, Draft; (4) the American Physical Society—"Report to the American Physical Society by the Study Group on Nuclear Fuel Cycles and Waste Management," New York, 1977 (reprinted in *Reviews of Modern Physics* 50, Suppl. [Jan. 1978]: S1).

Much additional useful information can be found in R. D. Lipschutz, *Radioactive Waste Politics, Technology, and Risk* (Cambridge, Mass.: Ballinger, 1980).

Finally, for examples of the rapidly growing technical literature on nuclear waste disposal we refer to the proceedings of the annual meetings of the Materials Research Society, published under the title *Scientific Basis for Nuclear Waste Disposal* (New York: Plenum Press, vol. 1, 1979; vol. 2, 1980; vol. 3, 1981).

2. A. Wolman and A. E. Gorman, "Waste Materials in the United States Atomic Energy Program," WASH–8, U.S. AEC, Jan. 1950. In this publication the following treatment of high-level radioactive waste was envisioned: "High-Level Radioactive Wastes. The irradiated products of reactors contain high-level radioactivity. These wastes have been called the ashes of the nuclear furnace but they cannot be disposed of as ordinary ashes. The most highly radioactive wastes are those remaining after the product desired, such as plutonium, is removed from the irradiated fuel by chemical separation processes. These wastes contain various fission products and inadequately irradiated uranium. They are highly dangerous because of their radioactivity. They are extremely valuable because of the recoverable uranium and other important materials they contain. Currently, these highly dangerous wastes are stored, but ways and means of recovering the valuable products in them are subject to much investigation. After certain cycles of decontamination have been completed the level of radioactive contamination drops very materially. As in the case of other wastes, a point is finally reached at which a decision must be made between the economics of further decontamination and the realities as to public health risks involved in release of these wastes to nature. Under present circumstances prudence dictates a conservative course of action in favor of protection of public health" (p. 9).

3. National Academy of Sciences—National Research Council, "The Disposal of Radioactive Waste on Land" (Washington, D.C., 1957).

4. U.S. AEC Authorizing Legislation, Fiscal Year 1972, Hearing before the Joint Committee on Atomic Energy, part 3, March 16–17, 1971.

5. J. E. Mendel and I. M. Warner, "Temperature Effects on Leaching Rate," in Battelle Pacific Northwest Laboratories Quarterly Progress Report BNW–1761, edited by A. M. Platt, June 1973, p. 4.

6. G. J. McCarthy, W. B. White, R. Roy, B. E. Scheetz, S. Komarneni, D. K. Smith, and D. M. Roy, "Interactions Between Nuclear Waste and Surrounding Rock," *Nature* 273 (1978): 216. See also A. E. Ringwood, S. E. Kesson, N. E. Ware, W. O. Hibberson, and A. Major, "The SYNROC Process: A Geochemical Approach to Nuclear Waste Immobilization," *Geochemical Journal* 13 (1979): 141.

7. G. J. McCarthy, "High-Level Waste Ceramics: Materials Considerations, Process Simulation and Product Characterization," *Nuclear Technology* 32 (1977): 92.

8. Ringwood et al., "The SYNROC Process."

9. See *Scientific Basis for Nuclear Waste Disposal,* in particular vol. 3.

10. U.S. AEC Authorizing Legislation, Fiscal Year 1972.

11. U.S. Nuclear Regulatory Commission, "Technical Criteria for Regulating Geologic Disposal of High-Level Radioactive Waste," *U.S. Federal Register* 45, no. 94 (May 13, 1980): 32400.

12. T. E. Scott, "Materials for Nuclear Waste Disposal Canisters: A Perspective," *Transactions, American Nuclear Society Annual Meeting,* Las Vegas, Nev., June 9–12, 1980, p. 196; L. Abrego and J. W. Braithwaite, "Corrosion of Titanium in Saline Nuclear Waste Isolation Environments," in ibid., p. 196; R. E. Westerman, "Investigation of Metallic, Ceramic, and

Polymeric Materials for Engineered Barrier Applications in Nuclear-Waste Packages," Battelle Pacific Northwest Laboratories Report no. PNL–3484, Oct. 1980.

3. J. A. Apps and N. G. W. Cook, "Backfill Barriers: The Use of Engineered Barriers Based on Geologic Materials to Assure Isolation of Radioactive Wastes in a Repository," in *Scientific Basis for Nuclear Waste Management,* edited by J. G. Moore (New York: Plenum Press, 1981), 3:291.

4. See NAS-NRC, "The Disposal of Radioactive Waste on Land."

5. Evidence for the official recognition of these problems can also be found in the fact that during the past three years the federal government's assumed opening date for a high-level waste depository has been moved back fully nine years, from 1988 to 1997, while in the Federal Republic of Germany the decision about the suitability of a single salt dome (at Gorleben) is expected to require twelve more years of investigation. See Michael Knapik, "DOE Says Carter's Waste Strategy Shift Has Doubled the Cost of Waste Disposal," *Nuclear Fuel* 5, no. 25 (Dec. 8, 1980): 12; *Nuclear Fuel* 6, no. 8 (April 13, 1981): 13.

6. P. R. Dawson and J. R. Tillerson, "Salt Motion Following Nuclear Waste Disposal," *Proceedings of the International Conference on Evaluation and Prediction of Subsidence,* Pensacola, Fla., Jan. 15–20, 1977 (report no. CONF. 780136–1).

7. W. D. Weart, "WIPP: A Bedded Salt Repository for Defense Radioactive Waste in Southeastern New Mexico," in *Radioactive Waste in Geologic Storage,* edited by Sherman Fried; ACS Symposium, Miami Beach, Fla., Sept. 11–15, 1978 (ACS Symposium Series 100; Washington, D.C.: American Chemical Society, 1979), pp. 14, 32.

8. See note 1, above, reference (1). In the foreword to the Geological Survey Circular 779, W. A. Radlinski, acting director of the U.S. Geological Survey, summarized his judgment as follows: "The many weaknesses in geologic knowledge noted in this report warrant a conservative approach to the development of geologic repositories in any medium. Increased participation in this problem by earth scientists of various disciplines appears necessary before final decisions are made to use repositories."

9. According to a report to the American Physical Society: "When we consider the relative likelihood of biospheric contact with geologically buried plutonium as opposed to piled Ra-226, the mill tailings may well be more important for the long term" (see note 1, above, reference [1], p. S-19). The Interagency Review Group on Nuclear Waste Management (IRG) stated the need for careful management of these wastes as follows: "By virtue of their presence at the surface, the actinide elements in mill tailings may constitute a greater potential problem than those in deeply buried HLW and TRU [high-level and transuranic] wastes. Thus, disposal of these tailings must be managed as carefully as that for HLW and TRU wastes" (see note 1, above, reference [3b], p. 227).

20. E. Landa, "Isolation of Uranium Mill Tailings and Their Component Radionuclides from the Biosphere—Some Earth Science Perspectives," U.S. Geological Survey Circular no. 814, 1980.

21. Natural uranium consists of two isotopes, the fissile U-235 and the more abundant (99.3 percent) U-238. They are both unstable, and decay with half-lives of 0.71 and 4.5 billion years, respectively, to nonradioactive lead. This decay occurs in a number of steps, through successive emissions of ionizing radiation (alpha and beta particles, and gamma rays). Because of this stepwise decay, every piece of uranium ore will contain traces of elements of the decay series, like thorium, radium, radon, and so on.

22. BEIR Report, *The Effects on Population of Exposure to Low Levels of Ionizing Radiation,* report of the Advisory Committee on the Biological Effects of Ionizing Radiation (Washington, D.C.: National Academy of Sciences, 1972).

23. Landa, "Isolation of Uranium Mill Tailings."

24. J. A. Adams and V. C. Rogers, "A Classification System for Radioactive Waste Disposal— What Goes Where?" U.S. NRC Report NUREG–0456, June 1978, p. 140.

25. J. J. Swift, J. M. Hardin, and H. W. Calley, "Potential Radiological Impact of Airborne Releases and Gamma Radiation to Individuals Living near Inactive Uranium Mill Tailings Piles," Report EPA–420/1–76–001, U.S. EPA, Office of Radiation Programs, Jan. 1976.

26. Michael Knapik, "National Institute for Occupational Safety and Health (NIOSH) to Lead Interagency Effort to Reduce Limits for Exposures of Miners to Radiation," *Nuclear Fuel* 5, no. 24 (1980): 10.

27. "Engineering Assessment of Inactive Mill Tailings, Mexican Hat Site, Utah," Ford Bacon and Davis Utah, Inc., Phase II, Title I, 1977.

28. Cited in U.S. Nuclear Regulatory Commission, Final Generic Environmental Impact Statement on Uranium Milling, NUREG–0706, Sept. 1980, p. A-36.

29. Landa, "Isolation of Uranium Mill Tailings."

30. Ibid.

31. "Money Spent for Clean-up Cost Estimates Would Have Been Enough to Cover Clean-up," *Nuclear Fuel* 5 (July 21, 1980): 9; "NRC Takes Charge of Edgemont Clean-up But Its Authority To Do So Is Unclear," *Nuclear Fuel* 5 (Aug. 18, 1980): 4.

32. R. H. Kennedy, "Long Term Stabilization of Uranium Mill Tailings at Inactive Sites," in *Tailings Disposal Today,* edited by G. O. Argall, vol. 2, Proceedings of the Second International Symposium, Denver, May 10, 1978; "Eldorado Nuclear Spells Out Its Need To Get Okay for New UF_6 Refining Capacities," *Nuclear Fuel* 5 (Sept. 20, 1980): 1; "AECB Weighing Whether To Continue Clean-up of Port Hope Radium Wastes," in ibid., Dec. 22, 1980, p. 13.

33. Landa, "Isolation of Uranium Mill Tailings."

34. U.S. Department of Energy, Progress Report on the Grand Junction Uranium Mill Tailings Remedial Action Program, DOE/EV–0033, Feb. 1979.

35. This rather substantial contamination was produced by a relatively small amount of mill tailings: Only about 50,000 tons of tailings were used as foundation material containing an estimated 40 curies of radium. The uranium extracted from these tailings would have been enough to operate one 1000-megawatt electric nuclear power plant for only about six months. Fifty thousand tons is only 0.03 percent of the present inventory of all mill tailings in the U.S. (estimated to be 150 million metric tons, according to K. K. S. Pillay, "An Independent Overview of Technology Development for Nuclear Waste Processing," *American Nuclear Society Transactions,* 1980 Annual Meeting, Las Vegas, Nev., June 9–12, 1980, p. 421). It is also interesting to note that, per unit energy produced, the amount of mill tailings equals that of the ash produced in a coal-burning power plant, assuming 0.1 percent uranium in the ore, and 10 percent ash in the coal. Hence the claim which is often made that the nuclear fuel cycle produces only small amounts of waste is incorrect.

36. Landa, "Isolation of Uranium Mill Tailings."

37. The NRC has proposed (see U.S. Nuclear Regulatory Commission, "Uranium Mill Licensing Requirements," Final Rules, *Federal Register* 45, no. 194 [Oct. 3, 1980]: 65521–38), over the vigorous objection of the mining industry which fears the additional expense (see "Kerr-McGee Files Suit to Overturn NRC's New Uranium Mill Licensing Rules," *Nuclear Fuel* 5, no. 22 [Oct. 27, 1980]: 4; "Kennedy Spurns American Mining Council's Attempt To Stall Mill Licensing Regulations at NRC," in ibid., no. 25 [Dec. 8, 1980]: 16), that future mill tailings should preferably be buried below the ground surface, but above the groundwater table. They should be covered with soil such that the atmospheric release rate of the radon would be reduced to 2 picocuries per square meter per second, which is twice the average radon release rate of soil. In cases where burial below grade would not be feasible, above-grade disposal and a soil cover might also be permissible (See U.S. NRC, Final Rules).

   While these measures would be improvements over the present practice, there are several reasons why they would still be inadequate. First, no soil or rock cover can be relied on to last hundreds, or even thousands of years. The mill tailings, however, will lose only one-half of their toxicity in 80,000 years (the half-life of the parent isotope thorium-230). For further protection the NRC has suggested that there be continued surveillance of the disposal sites; this is even more likely to fail on the time scales involved. Second, burial below grade will reduce the distance between tailings and groundwater. Radium leached from the tailings would be a serious contaminant for the latter. In the extreme case the groundwater table could rise into the mill tailings pile (perhaps as a result of a change in climate), and people would withdraw drinking water from it. Resulting exposure rates for individuals have apparently not yet been estimated; if we consider that the 80 curies of radium contained in the mill tailings produced in providing the fuel for operating a large reactor for one year could contaminate eight billion cubic meters of water (the average annual amount of water carried by a good-size stream) to the maximum permissible concentration for the general population (called MPC_w; see U.S. National Bureau of

Standards, "Maximum Permissible Body Burdens and Maximum Permissible Concentrations of Radionuclides in Air and Water for Occupational Exposure," NBS Handbook 69, June 1959), we expect the dose rates in the well scenario to be very high. In order to reduce the risks of groundwater pollution, the NRC has recommended the use of liners made of clay or plastic. Their lifetimes are at least as doubtful as those of the covers.

38. Landa, "Isolation of Uranium Mill Tailings."

39. See U.S. NRC, Final Generic Environmental Impact Statement.

40. D. S. Metlay, "History and Interpretation of Radioactive Waste Management in the United States," in *Essays on Issues Relevant to the Regulation of Radioactive Waste Management,* edited by W. P. Bishop (Washington, D.C.: U.S. NRC, NUREG 0412, 1978).

41. Battelle Pacific Northwest Laboratory, "Reference Site Initial Assessment for a Salt Dome Repository," Report PNL–2955, Dec. 1979.

42. BEIR Report, 1972.

43. It is often argued that after 1000 years the nuclear wastes are no more toxic than the uranium ore from which the fuel was produced; consequently there would be little reason to worry about the toxicity of the nuclear waste beyond that time. The fallacy of this argument is that the unmined uranium ore is well locked up in its natural geologic setting, where it has been for millions of years, while the nuclear waste in a repository is in a highly artificial form and in an unnatural setting. The same is to be said about the leftovers from the uranium mining, the mill tailings, as we saw earlier. Furthermore, the repository will contain the wastes resulting from a hundred reactors operating for several decades, in a single place covering an area of a few square miles; the ore from which the corresponding fuel was made, however, came from many different ore bodies scattered over hundreds of thousands of square miles.

44. K. S. Johnson and S. Gonzales, "Salt Deposits in the United States and Regional Geologic Characteristics Important for Storage of Radioactive Waste," Office of Waste Isolation Report Y/OWI/SUB–7414/1, March 1978, p. 174.

45. In light of these considerations, the following recommendations, made recently by the National Academy of Sciences, Committee on Radioactive Waste Management, Panel on Geologic Site Criteria ("Geological Criteria for Repositories for High-Level Radioactive Wastes," Washington, D.C., Aug. 1978) make a great deal of sense: "No area with a present or past record of resource extraction, other than for bulk materials won by surface quarrying, should be considered as a geological site for radioactive wastes. This restriction rests on one or more of three possible considerations: (a) present or predictable future importance as a potential source of needed raw materials; (b) disturbance of the natural hydrologic regime in consequence of present or past underground development and exploration, such as tunneling, hydraulic fracturing, etc., resulting in greater uncertainty as to the paths and volumes of fluid flow; and (c) potential attractiveness to future developers and explorers for natural resources who may be drawn to the area by evidence of past activities of resource extraction."

If we further consider the high water solubility of salt and the corrosiveness of the resulting brine, we cannot help but conclude that, quite apart from any geologic considerations, salt formations must be judged unsuitable for the disposal of nuclear wastes (or, in fact, any highly toxic wastes). Based on the same argument, formations of the abundant silicate rocks like granite, gneiss, tuff, or basalt should be more acceptable, in particular if they occur at great depth, making them far less attractive for mining than the same material close to the surface.

46. U.S. NRC, "Technical Criteria," p. 31396.

47. Robert Gillette, "Radiation Spill at Hanford: Anatomy of an Accident," *Science* 181 (1973): 728.

48. Lipschutz, *Radioactive Waste.*

49. Ibid. See also, *Nucleonics Week* 21, no. 35 (Aug. 28, 1980): 11.

50. Z. A. Medvedev, *Nuclear Disaster in the Urals* (New York: W. W. Norton, 1979); J. R. Trabalka, L. D. Eyman, and S. I. Auerbach, "Analysis of the 1957–1958 Soviet Nuclear Accident," *Science* 209 (1980): 345.

Chapter 5. Economics

1. Washington Analysis Corporation, "Nuclear Energy: Dark Outlook," Dec. 21, 1979, p. 1.

2. H. E. Van, M. J. Whitman, and H. I. Bowers, "Factors Affecting the Historical and Projected

Capital Costs of Nuclear Plants in the U.S.A.,'' *Proceedings,* Fourth International Conference on the Peaceful Applications of Atomic Energy, vol. 2 (Geneva, Sept. 1971), pp. 21–43.

3. Saunders Miller, *The Economics of Nuclear and Coal Power* (New York: Praeger, 1976), p. 109.

4. Kathryn M. Welling, ''Energy Alternatives: Blueprint for the Future,'' *Barron's,* May 30, 1977.

5. ''Nuclear Energy: Struggling Against Murphy's Law,'' *Chemical and Engineering News,* March 7, 1977, p. 8; ''Nuclear Power Costs,'' Twenty-Third Report by the House Committee on Government Operations, U.S. House of Representatives, 1978, p. 2; and Atomic Industrial Forum INFO news release, Jan. 17, 1979.

6. ''Nuclear Power Costs,'' p. 31.

7. Ibid.

8. Charles Komanoff, ''Cost Escalation at Nuclear and Coal Power Plants,'' Komanoff Energy Associates, New York, 10016, Aug. 15, 1980, p. 17.

9. Robert Stobaugh and Daniel Yergin, *Energy Future: Report of the Energy Project at the Harvard Business School* (New York: Random House, 1979), p. 121.

10. Marc Messing, ''Reasons for Delay in Powerplant Licensing and Construction: An Initial Review of Data Available on Powerplants Brought on Line from 1967 Through 1976,'' Environmental Policy Institute, Jan. 1978, p. 1.

11. Federal Power Commission Utility questionnaires to private, investor-owned utility companies, 1974 (Public Document Room, U.S. Department of Energy, Washington, D.C.).

12. William O. Doub, *Nuclear Powerplant Siting & Licensing,* Hearings before the Joint Committee on Atomic Energy, 93rd Congress, March 19, 1972, pp. 9–10.

13. Messing, ''Reasons for Delay,'' p. 6.

14. Report to the Congress by the Comptroller-General of the United States, ''Reducing Nuclear Powerplant Leadtimes: Many Obstacles Remain,'' March 1977, p. 5.

15. Ibid., p. iii.

16. The Library of Congress, Congressional Research Service, ''The Role of Licensing in Nuclear Plant Construction Times,'' by Carl E. Behrens, Analyst, Environment and Natural Resources Policy Division, Oct. 20, 1977, pp. 1–2, 10.

17. ''Programmatic Information on the Licensing of Standardized Nuclear Powerplants,'' U.S. Atomic Energy Commission, Aug. 1974.

18. Nuclear Regulatory Commission, Special Inquiry Group, ''Three Mile Island: A Report to the Commissioners and To the Public,'' Jan. 1980, 1:139.

19. Ibid.

20. Ibid., p. 140.

21. Stobaugh and Yergin, *Energy Future,* p. 119.

22. Charles Komanoff, ''A Comparison of Nuclear and Coal Costs,'' testimony before the State of New Jersey, Brand of Public Utilities, Phase III, Oct. 9, 1978.

23. Charles Komanoff, *Power Plant Performance* (New York: Council on Economic Priorities, 1976).

24. David Burnham, ''GE Warns of Halt in Making Reactors,'' *New York Times,* May 15, 1977, p. 1.

25. ''Public Utilities Review,'' Moseley, Hallgarten, Estabrook & Weeden, Inc., April 1979, p. 2.

26. Quoted in *Barron's,* Aug. 24, 1981, p. 21.

27. ''Report of the President's Commission on the Accident at Three Mile Island,'' Oct. 1979, p. 19.

28. Gulf States Utility Co. (River Bend Units 1 and 2), ALAB–183, RAI–74–3, decided March 1, 1974, Slip Op., pp. 10–12.

29. *Nuclear Reactor Safety,* Part 1, Hearings, Joint Committee on Atomic Energy, U.S. Congress, Jan., Sept., Oct., 1973, p. 51, as cited by Ralph Nader and John Abbotts in testimony before the House Subcommittee on Energy and Environment, June 6, 1978, p. 19.

30. NRC, ''Report to the Commissioners,'' 1: 143–44.

31. Ibid., p. 143.

32. ''Distributed Energy Systems in California's Future: Final Report,'' Prepared for the U.S. Department of Energy, Oct. 1978. The study, a joint project of the Lawrence Berkeley Laboratory, the Lawrence Livermore Laboratory, and DOE discovered that the state of California could rely solely on renewable energy technologies by the year 2025 for all its energy needs despite continued economic and population growth.

33. Jack Horan, ''Three Mile Island Could Generate a $7 Billion Bill,'' *Charlotte Observer,* March 16, 1980. The $7-billion cost relates to known financial losses due to the decontamination of the damaged TMI Unit 2, the loss of

reactors for safety checking, and "secondary costs" related to federal activities and new regulatory requirements. Since this estimate was written the Metropolitan Edison Company sharply escalated its decontamination costs and suffered additional losses in rate revenues due to actions by the Pennsylvania Public Utilities Commission.

34. Joanne Omang, "TMI's Owners Struggle with Radioactive and Financial Fallout," *Washington Post,* March 19, 1981.

35. "Questions on the Future of Nuclear Power: Implications and Trade-offs," Comptroller-General's Report to the Congress, U.S. General Accounting Office, May 21, 1979.

36. Ibid.

37. "The Impact of Regulation on Utilities," speech by Leonard S. Hyman before the Pennsylvania Power Conference, April 30, 1980.

38. "Appraising Regulation," by Eunice T. Reich, seminar on utility finance for outside directors, Sept. 17, 1980.

39. "The Future of Nuclear Power," speech by Leonard S. Hyman before the American Nuclear Society, Windsor, Conn., Jan. 14, 1980.

40. Ibid.

41. Pat Choate, "As Time Goes By—The Costs and Consequences of Delay," The Academy for Contemporary Problems, May 12, 1980.

42. These estimates coincide with a report by the U.S. General Accounting Office which projects that the national demand for electricity could require electrical generating capacity to more than double by the end of the century. See "Electricity Planning—Today's Improvements Can Alter Tomorrow's Investment Decisions," Comptroller-General's Report to the Congress, U.S. General Accounting Office, Sept. 30, 1980.

43. "The Future of Nuclear Power," speech by Leonard S. Hyman.

44. Data collected and published by the Atomic Industrial Forum, the major public relations and lobbying organization of the nuclear industry, shows electricity from nuclear plants beginning operation after 1975 to be more expensive than from coal-fired plants completed in the same years: "U.S. Average Electrical Generating Costs and Capacity Factors in 1979," *INFO Special Report,* Atomic Industrial Forum, Feb. 1981.

45. Unless otherwise noted, all calculations herein are by the author, based on data in various issues of the *Monthly Energy Review* and *Electric Power Monthly,* Department of Energy, Energy Information Agency, Washington, D.C.

46. Calculation by the author. Current emissions of pollutants are assumed to be approximately equal to those reported for 1975 in *Staff Findings,* The President's Commission on Coal, March 1980, Table 1, p. 27. Generation of electricity by type of fuel is from the *Monthly Energy Review,* note 45.

47. Fuel savings, for a plant operating at 50 percent of theoretical capacity, would equal $130 per kilowatt per year. A scrubber on a new plant is estimated to cost $135 per kilowatt. Allowing a cost penalty of one-half to account for the greater difficulty of retrofitting a scrubber, its cost would be recovered by fuel savings from the oil to coal conversion in 1.5 years. The total cost of improved air-pollution controls, including installation of an electrostatic precipitator with at least 99.5 percent efficiency, would be recovered in about two and a half years. Estimates of the cost of air-pollution controls are in 1980 dollars and are from Charles Komanoff, "Pollution Control Improvements in Coal-Fired Generating Plants: What They Accomplish, What They Cost," *Air Pollution Control Association Journal* 30, no. 9 (Sept. 1980).

48. The rate of rise would be about 3 percent per eleven years instead of per ten years. This result is based on the following calculation: If fossil energy consumption grows at 2 percent per year until 2000, and the share of various fuels remains as it is today, annual production of carbon dioxide would equal 8 billion tons. If 500,000 megawatts of nuclear power displaced an equivalent amount of coal, it would reduce carbon dioxide production by 600 billion tons, or 7.5 percent; thus it would take 7.5 percent longer to add the same amount of CO_2 to the atmosphere. Source: Gordon J. F. MacDonald, "An Overview of the Impact of Carbon Dioxide on Climate," The Mitre Corp., Arlington, Va., Dec. 1978. For the estimate of nuclear power growth, see Vince Taylor, *Energy: The Easy Path* (Cambridge, Mass.: Union of Concerned Scientists, Jan. 1979).

49. Stobaugh and Yergin, *Energy Future.*

50. R. Sant, *The Least Cost Energy Strategy* (Arlington, Va.: Mellon Institute, 1979); and R. Sant and S. Carhart, *Eight Great Energy*

Myths (Arlington, Va.: Mellon Institute, 1981).

51. "Low Energy Futures for the United States," U.S. Dept. of Energy, DOE/PE–0020, June 1980.

52. For example, "U.S. Energy Demand: Some Low Energy Futures," *Science* 200, no. 4338 (April 14, 1978), presents a summary of the findings of the Demand and Conservation Panel on the National Research Council's Committee on Nuclear and Alternative Energy Systems (CONAES); Amory Lovins et al., "Nuclear Power and Nuclear Bombs," *Foreign Affairs* (Summer 1980), present an overview of recent findings on the potential of conservation; both empirical and theoretical reasons why improvements in energy efficiency are preferable to efforts to expand supply are presented in Taylor, *Energy: The Easy Path,* and id., *The Easy Path Energy Plan* (Cambridge, Mass.: Union of Concerned Scientists, Sept. 1979).

53. A reasonably detailed analysis showed that every area of the country would have been able to fulfill its peak electrical demands in 1979 in the absence of nuclear power (R. Carlson, D. Freedman, and R. Scott, "A Strategy for a Non-Nuclear Future," *Environment,* July/Aug. 1979). This was possible because of a combination of substantial excess generating capacity and extensive transmission interconnections. As of 1981 the situation is about the same as in 1979; but by 1985, when generating surpluses seem likely to be much smaller and, on present plans, nuclear power will have come to provide a larger proportion of electricity supply, a shutdown will not be so easily accommodated. This, however, hardly provides an argument *for* more nuclear power.

54. *Non*nuclear generating capability was 501,000 megawatts, 17 percent greater than the peak demand of 428,000 megawatts, a margin generally considered adequate to provide reliable service—although shortages might have occurred in some areas in the absence of nuclear generation (see note 53). Sources: *Monthly Energy Review* and *Electric Power Monthly,* note 45.

55. More than half of 1980 nuclear generation appeared replaceable by coal rather than oil (Vince Taylor, "Electric Utilities, A Time of Transition," *Environment,* May 1981).

56. See, note 1 and also Charles Komanoff, "Power Plant Cost Escalation," Komanoff Energy Associates, New York City, 1981. He concludes, after analyzing trends in costs of all nuclear and coal plants completed in the United States between 1972 and 1978, that for plants coming into operation in the late 1980s nuclear electricity will be 20–25 percent costlier than coal-generated electricity.

57. Operating at 60 percent of capacity, 100,000 megawatts of nuclear power would generate 526 billion kilowatt-hours of electricity. In 1980 the U.S. consumed 260 billion gallons of petroleum products.

58. W. Ramsey, "Coal and Nuclear: Health and Environmental Costs," draft report prepared for National Energy Strategies Project, Washington, D.C., 1978 (available from Resources for the Future's Center for Energy Policy Research), pp. 1–31 and 1–32.

59. S. H. Schurr et al., *Energy in America's Future* (Baltimore: Johns Hopkins University Press, 1979), p. 346.

60. Panel on Energy and Climate, National Academy of Sciences, "Energy and Climate," Washington, D.C., 1977.

61. Our elasticity estimate is identical to that employed in the Energy Modeling Forum's World Oil Study. That is, a 1 percent increase in "constant dollar" primary energy prices leads eventually to a decrease of 0.4 percent in demands. This estimate is based on an extensive review of domestic and international experience with conservation since the first price shocks in 1973. See Energy Modeling Forum, "World Oil—EMF 6 Summary Report," Stanford University, Stanford, Calif., Sept. 1981.

62. Schurr, *Energy in America's Future,* p. 322.

63. Ibid., p. 300.

64. A. S. Manne et al., "ETA-MACRO: A User's Guide," EA–1724, Electric Power Research Institute, Feb. 1981.

65. For example, see S. M. Keeny et al., *Nuclear Power Issues and Choices,* Report of the Nuclear Energy Study Group (NEPS) (Cambridge, Mass.: Ballinger, 1977); and Modeling Resource Group (MRG), "Energy Modeling for an Uncertain Future," Committee on Nuclear and Alternative Energy Systems (CONAES), National Research Council, National Academy of Sciences, Washington, D.C., 1978.

66. In addition the model accounts for oil- and gas-fired electricity and for hydroelectric and geothermal. In 1980 hydroelectric, geothermal, etc., accounted for 12 percent of the electricity generated. It is assumed that these sources will

expand at the annual rate of 3 percent in the future. Other detailed technical assumptions are documented in an appendix that is available upon request.

67. For coal-fired units and LWRs, these costs have been adapted from the most recent estimates available at the Electric Power Research Institute. See Electric Power Research Institute, "Technical Assessment Guide," Draft, Palo Alto, Calif., Aug. 1981.

68. For previous estimates of these overall impacts, see Institute for Energy Analysis, *Economic and Environmental Impacts of a U.S. Nuclear Moratorium, 1985–2010* (Cambridge, Mass.: MIT Press, 1979); Keeny, "Nuclear Power Issues and Choices"; and Modeling Resource Group, "Energy Modeling for an Uncertain Future."

Index

accidents, nuclear plant:
  in breeder reactors, 175–76, 211–12
  at Browns Ferry reactor, 91–92
  cancer death risk in, 51–52, 65, 74, 102,
    103, 224–25
  Class Nine, 222, 224–25, 226
  consequences of, 88–89, 97–98
  criticality (reactivity), 21, 175–76
  at Fermi I reactor, 21
  in fusion reactors, 199, 205
  as human errors, 26
  loss-of-coolant, 22, 81–82, 87–88, 92,
    199, 221
  in meltdown scenario, 98–102
  other risks vs., 89, 221
  probability of, 88–89, 222
  Rasmussen report on, 24, 89–90
  see also Three Mile Island nuclear plant
    accident
agriculture, radiation used in, 53
Albert, R. E., 68
Alcator A tokamak, 204
Aldermaston Peace Marchers, 237
alpha rays, 29, 36, 64n
aluminum smelters, 181
American Cancer Society, 129
American Nuclear Society, 172
American Physical Society, 24, 112
antinuclear movement, 148, 219, 233–38,
    240–43
  government deceptions as spur to, 234–35
  larger social problems addressed by, 236,
    242–43
  in mass actions, 237
  other energy forms opposed by, 240–42
  nuclear victims in, 236
antiwar movement, 236, 237

Argentina, 229
As Low As Reasonably Achievable
    (ALARA) exposure standard, 42–43,
    45, 50
Atomic Bomb Casualty Commission, 28
atomic bombs, 15–18, 213, 233
Atomic Energy Act (1946), 18, 19, 228–29
Atomic Energy Authority, United Kingdom,
    62
Atomic Energy Commission (AEC), 18, 19,
    27, 44, 105
  accidents studied by, 82
  deaths unconceded by, 61–62
  disinformation generated by, 60–61
  Environmental Impact Statements pre-
    pared by, 144
  Hanford study funded by, 33, 54–55
  "hot particle" study by, 77
  improbability of accidents argued by, 86,
    89
  military research promoted by, 18
  nuclear costs projections by, 142
  on nuclear waste, 124–25
  on public participation, 147
  reactor safety research funded by, 83
  replacement of, 24, 248–49
  Shaw's management of, 20, 21, 22–23
  sit-in at, 234
  in waste disposal projects, 125, 126
Atomic Industrial Forum (AIF), 148–49, 172
atomic workers:
  "burning out" of, 42, 45
  cancer among, 33, 40–41, 45
  occupational safety standards for, 31, 35,
    45, 49–50
  proposal for training of, 96
  radiation exposure of, 33, 40–41, 42–43

atoms, 15–16
Atoms for Peace program, 18, 44, 204, 229, 234
automobiles:
  electric, 223–24
  fuel conservation in, 181, 186–87
Avery, Robert, 175–76, 206–16

Babcock and Wilcox, 20, 145
Bank of America, 145
Barnwell reprocessing plant, 109, 237
Barron's, 141, 146
Baruch Plan, 228
Battelle Pacific Northwest Laboratories, 126, 130
Becquerel, Antoine Henri, 36
benefit-risk doctrine, 58–59, 68
beta rays, 29, 64n
Bethe, Hans, 175–76, 206–16
Beyea, Jan, 97–107
Bikini Atoll, 204
Biological Effects of Ionizing Radiation (BEIR) Committee reports, 51, 61–63, 65, 66, 71, 78, 130, 225, 254–55n
biomass conversion, 174, 241
boiling-water reactors (BWRs), 22, 88, 100, 106, 145
Borax I reactor, 100
Brazil, 229, 231, 232
breeder reactors, 171, 175–76, 206–16
  criticality accidents in, 175–76
  deployment of, 213–14
  disassembly accidents in, 212
  economics of, 214, 221
  fuel supplies stretched by, 20, 175, 206–7
  government position on, 175, 212–13, 215, 216
  light-water reactors vs., 164, 207–11, 214
  objections to, 175, 212–13, 230
  plutonium in, 175, 207, 208, 209, 210, 212–13, 230
  pool vs. loop design in, 211
  research on, 20–21
  in Second Nuclear Era, 220
  unlimited energy produced by, 206–7
  waste disposal in, 175, 207–8

British x-ray patient study, 28, 29, 34, 37, 41, 62
Britt, Russell, 149
Brodsky, Allen, 46–56
Bross, Irwin, 39
Browns Ferry Nuclear Power Station, 24, 25, 91–92, 106
Bupp, Irvin C., 140, 153, 248

California referendum on nuclear plant construction (1976), 250
Calvo, Victor, 241
Canada, 16, 19, 176, 178, 182, 190
cancer, radiation-caused, 61–68
  absolute risk estimates of, 63–64
  age factor in, 64–65, 253–55n
  among atomic workers, 33, 40–41, 45
  bronchogenic, 67–68, 256n
  cancer-dose calculations of, 64–65
  cell reproduction in, 29
  chance nature of, 48–49
  delay in appearance of, 28, 90, 91
  diminishing effect hypothesis for risk of, 66
  government studies on, 61–63
  among Hiroshima-Nagasaki victims, 36–37, 41, 51, 54, 66, 72
  leukemia, 37, 39, 41, 50, 76
  levels of radiation in, 72
  life-shortening results of, 52
  linear hypothesis for risk of, 66, 71–72, 224–25, 251n
  of lungs, 36, 43, 50–51, 67, 78, 128–29
  from meltdown accident, 102, 103
  among miners, 36, 43, 50, 67
  myeloma, 40, 41, 54
  O/E function in, 63–64
  overestimation of risks of, 71–72
  from oxygen, 225
  from plutonium exposure, 43, 256n
  from prenatal exposure, 37–38, 39, 225
  radiation treatments for, 53
  among radiologists, 50
  from radon gas exposure, 128–29
  relative risk estimates of, 63
  supralinearity hypothesis for risk of, 66, 251n

among test site residents, 76
from Three Mile Island accident, 51–52, 65, 74
underestimation of risk of, 64–65
from vented gases, 42
carbon dioxide, 158, 162
carbon dioxide-cooled reactors, 19
Carter administration, 109, 148, 193, 204
breeder reactor project deferred by, 175, 212, 215
nonproliferation policies of, 227, 230, 232
Center for Fusion Engineering (CFE), 204
Center for Science in the Public Interest, 141
Chadwick, James, 15
Chaim Sheba Medical Center, 39–40
Chicago 8, 233
China, People's Republic of, 227, 229, 231
China Syndrome, 88, 221
citizen groups, 146–47
Clamshell Alliance, 142
Clark, William, 225–26
Class Nine accidents, 222, 224–25, 226
Clinch River Breeder Reactor, 138, 175, 215
coal:
as electricity source, 138, 139, 140, 156, 161
health hazards produced by, 32, 42, 75, 76, 90, 107, 162, 257n
in nuclear phase-out scenario, 165–70
nuclear power costs vs. costs of, 23, 142, 149, 155, 160, 161, 180
as oil substitute, 157, 158
pollution from, 158, 161–62, 180
supplies of, 165–66, 171
cobalt mines, 36
Cohen, Bernard L., 69–79, 85, 218–19
Combustion Engineering, 20, 145
Committee for Nuclear Responsibility, 57
Committee on Nuclear and Alternative Energy Systems (CONAES), 161
Commoner, Barry, 243
Congress, U.S., 146, 154, 156, 193, 200, 204
conservation, energy, 159, 162, 180–82, 185, 186–87, 223
control rods, reactor core, 21, 22, 87
cosmic rays, radiation in, 48
Council on Economic Priorities, 145

criminology, radiation used in, 53
criticality accidents, 21, 175–76
Critical Mass Energy Project, 141

Danforth, D. D., 153
Davis Besse reactor, 84, 222
Dawson, P. R., 127
Dean, Stephen O., 175, 198–206
decommissioning, reactor, 137, 146
"defense in depth" concept, 98
Dellinger, David, 219, 233–38
Denmark, 190
Detroit Edison, 20–21
deuterium, 198–99, 200, 203, 205, 206
Diablo Canyon nuclear plant, 26
dismantlement, as reactor decommissioning method, 137
DNA molecules, radiation damage to, 29, 72
Dolphin, G. W., 62
Domenici, Pete, 154
Donath, Fred A., 115–23
Doppler effect, 212
Doub, William O., 143

Edison Electric Institute, 27, 135, 137
Edwards, James B., 138, 154
Ehrlich, Paul, 242, 244
Einstein, Albert, 16, 17, 233, 238
Eisenhower administration, 18, 204, 229
Électricité de France, 187
electricity:
artificial demand for, 150
cheapest sources of, 185
conservation of, 159, 180–81, 185
cost comparisons for, 165
declining consumption of, 24, 136
growth predicted in consumption of, 152–53, 159
as high-quality energy, 184
non-nuclear sources of, 106–7, 138–40, 160
oil costs vs. costs for, 184
reactor production of, 22, 23, 106, 138, 148–49
rejection of, 223–24
Electric Power Research Institute (EPRI), 160, 204

elements, half-life of, 112–13
Eliot, T. S., 238
emergency core cooling system (ECCS), 23, 82, 87–88
    effectiveness of, 83, 84, 85
    failure of, 99, 221
    in Three Mile Island accident, 93
energy:
    cost-effectiveness and, 184–85
    efficiency improvements in use of, 180–82, 183
    least-cost strategy for, 182–83
    projected world demand for, 220
    "spaghetti chart" for flow of, 187–88
Energy and Climate Conference, 220
energy crisis, 25, 115, 138, 155, 160
Energy Department, U.S. (DOE), 41, 138, 147
    disinformation generated by, 60–61
    in fusion energy projects, 204
    solar-cell technologies anticipated by, 190
    waste studies by, 129
energy fairs, 171–72
Energy Future (Harvard Business School), 192, 223, 248
Energy Research and Development Administration (ERDA), 40–41
    disinformation generated by, 60–61
    establishment of, 24
    Gofman's work refuted by, 68
Energy: The Easy Path (Taylor), 155
entombment, as reactor decommissioning method, 137
environmental impact statements (EIS), 59, 143–44
Environmental Protection Agency (EPA), 71
    health effects estimates by, 99, 128–29
    nuclear waste studies by, 114
    pollution standards set by, 158
ETA-MACRO computer model, 163–64
ethyl alcohol, 194
Eve, I. S., 62
Experimental Breeder Reactor #2 (EBR-II), 215

Falk, Richard, 218, 226–33
Fast Flux Test Facility (FFTF), 215
Federal Power Commission (FPC), 142–43

Fermi, Enrico, 16, 175
Fermi I reactor, 21, 237
Finch, Robert, 61
fission, nuclear, 15–16
    control of, 26
    discovery of, 15–16
    fusion vs., 198, 199
Florida Power and Light, 137
Fonda, Jane, 243
Ford Foundation, 112
France, 180, 187–88, 190, 215, 216, 220, 227, 229
Franklin County Energy Project, 192
Freedom of Information Act, 41
Friends of the Earth, 243, 245
fuel rods, reactor, 102, 109, 112, 199
fusion energy, 162–63, 171, 174–75, 192, 198–206
    "break-even" point in, 199–200, 202, 203
    development of, 204–6
    energy problems solved by, 198–99
    environmental and safety advantages of, 199
    inertial confinement of, 202–4
    magnetic confinement of, 202
    plasma in, 200–202
    waste from, 199, 205
Fusion Engineering Device (FED), 204
Fusion Power Associates, 175

gamma rays, 29, 38, 48–49, 52, 64n
gases, radioactive, 28, 42, 48, 88, 199
gas industry, 150, 171
gasohol, 194
General Accounting Office (GAO), 143, 153
General Atomic Company, 145, 205
General Electric Company, 20, 84, 100, 145, 196, 235, 236
General Motors Corporation, 224
General Public Utilities Corporation, 137
genetic defects, radiation-caused, 28, 29, 32, 38–39, 48, 70, 71, 72–73
Geological Survey, U.S., 114
Germany, Federal Republic of, 176, 179, 190, 215, 231
Germany, Nazi, 16, 59
Gilinsky, Victor, 232–33
Gofman, John W., 32, 56–69, 218

Government Operations, U.S. House Committee on, 142
Grand Junction, tailings sand in, 129
Gravel, Mike, 61
Great Britain, 16, 19, 190, 204, 213, 215, 227
Gross National Product (GNP), energy and, 164, 183, 243
Grumman Aircraft Corporation, 196, 205
Guidebook to Nuclear Reactors, A (Nero), 86

Haefele, Wolf, 220
Hahn, Otto, 16
Hainesville Salt Dome, 131
Handbook of Radiation Measurement and Protection (Brodsky), 46
Handler, Philip, 136
Hanford Atomic Works, 33, 40, 53–55, 74, 111, 133
hard energy technologies, 189, 190
Harris polls, 149
Hart, Gary, 154–55
Harvard Business School, Energy Project of, 144, 159, 189, 192
Hayden, Tom, 243
Hayes, Dennis, 193
heating, fuel conservation in, 181–82, 185, 186
heat pumps, electric, 223–24
heavy-water reactors, 19, 220
Heidelberg report, 77
Hempelmann, Louis H., 67
Hertsgaard, Mark, 138
Hiroshima and Nagasaki, bombing of, 15, 28, 29, 33, 233, 234
  cancer deaths from, 36–37, 41, 51, 54, 66, 72
  data on, 36–37, 38, 51, 250
  genetic defects from, 73
Horn, Carl, Jr., 135
hot particle theory, 77–78, 256n
human rights, 57–58, 68–69
hydrogen bombs, 200, 234
hydropower, 171, 190, 193, 194
Hyman, Leonard S., 150, 151, 154, 155

India, 217, 227, 229, 231
Indian Point reactor, 85, 102

Indonesia, 231
induction process, 22
Industrial Revolution, 58
inflation, 148, 150, 151, 161
Institute for Energy Analysis, 222
Institute of Nuclear Power Operations (INPO), 96
International Atomic Energy Agency (IAEA), 229
International Commission on Radiological Protection (ICRP), 31, 35, 38, 67–68, 71
Iraq, 217
Israel, 217, 227, 229, 231

Japan, 161, 190, 215, 220, 231
Jersey Central Power and Light Company (JCPL), 20
Joint Committee on Atomic Energy (JCAE), 18–19, 20, 235

Kendall, Henry, 34, 81
Kennedy, Richard T., 241
Kerr-McGee Oklahoma fuel fabrication plant, 41
King, Martin Luther, 237
Kneale, George, 40, 41, 54, 225
Komanoff, Charles, 142, 145
Krulitz, Leo, 242

Lao-tzu, 192
Lapp, Ralph, 137
Lawrence Livermore National Laboratory, 38, 51, 56–57, 202, 203, 204
Leger, Gene, 181
Lewis, Floyd W., 32
Lewis, Harold W., 84, 85, 92
liberty, concept of, 57–58
Library of Congress, 143
light-water reactors (LWRs), 19–24, 207–11
  breeder reactors vs., 164, 207–11, 214
  core of, 22, 81
  design of, 104
  development of, 19–24
  fusion reactors vs., 199
  in nuclear submarines, 19
  operation of, 22
  safety research on, 20–23

light-water reactors (*continued*)
  safety systems in, 82
  scaling up of, 21
  uranium wasted by, 20, 44–45, 207
Lilienthal, David E., 247–48, 250
Linear Energy Transfer radiation, 64
liquefied natural gas (LNG), 189, 240
Liquid Metal Fast Breeder Reactor (LMFBR), 210, 211
Lisbon Declaration (1980), 231
Los Alamos Scientific Laboratory, 41, 67, 203–4
loss-of-coolant accidents (LOCAs), 22, 81–82, 87–88, 92, 199, 221
Loss of Fluid Test reactor (LOFT), 22–23, 83
Love Canal, 111
Lovins, Amory, 172–73, 176–92, 220, 242, 244
Lovins, Hunter, 172–73, 176–92
Lyon, Joseph, 76
Lyons waste disposal site, 124–27, 130, 133

McClue, James, 154
McCollam, William, 135
Magnetic Fusion Energy Engineering Act, 204
Maidique, Modesto A., 173–74, 192–98
Mancuso, Thomas, 33, 40–41, 46, 53–55
Manhattan Project, 16, 18, 67
Manne, Alan S., 160–70
Mazrui, Ali, 228
Medical Research Council (MRC), Great Britain, 77
meltdowns, 75–76, 82–83, 84–85, 86, 98–103
  cause of, 99
  containment buildings in, 100, 101
  environmental effects of, 101, 102
  evacuation during, 101–2
  fuel rods in, 102
  hypothetical nature of, 103
  probability of, 103–4, 105, 106, 221
  property claims after, 102–3
  scenario for, 98–102
  steam explosions in, 100–101
  "waves" of radiation from, 102
Menace of Atomic Energy, The (Nader), 141
Mendel, J. E., 126

Meredith, George, 234
Metropolitan Edison, 65
Milham, Samuel, 40
Miller, Saunders, 141, 142
Mirror Fusion Test Facility (MFTF), 202
Mobil Corporation, 155
Modan, B., 39–40
Morgan, Karl Z., 35–46
Moseley, Hallgarten, Estabrook & Weeden, 145–46
mothballing, as reactor decommissioning method, 137
Muller, H. J., 38–39

Nader, Ralph, 61, 75, 78, 103, 141–47, 243
National Academy of Sciences (NAS), 34, 51, 61, 71, 77, 112, 124
National Council on Radiation Protection and Measurement (NCRP), 31, 35, 71
National Institute for Occupational Safety and Health (NIOSH), 129
National Radiological Protection Board (NRPB), United Kingdom, 77
National Reactor Testing Station, 19, 82–83
Nation's Business, 148
natural reactors, 120
Natural Resources Defense Council (NRDC), 240–41
Nautilus, 19, 21
Nero, Anthony V., Jr., 86–97
neutron activation, 205
New York Times, 27, 75
Nichols, Edward, 241–42
Nixon, Richard, 175
Nonproliferation Treaty (NPT) (1968), 229–30, 232
Norton, Boyd, 15–26
Nova fusion machine, 203
nuclear industry:
  court interventions in, 137, 143
  "defense in depth" concept in, 98
  disinformation on, 59–60
  government support for, 19, 25, 57–59, 146
  public excluded from decision-making in, 143–44
  public relations campaign mounted by, 104, 155–56

radiation effects downplayed by, 44
Reagan administration support for, 137–38
regulatory climate for, 136, 138, 143, 149–51
rights of, 58
safety agencies established by, 96
safety record of, 84
nuclear power:
  Canadian projections for, 176, 178
  clustered sites for, 222–23
  commercial development of, 18–19
  costs of, 142, 147, 184–85, 186
  decline in need for, 136
  as dispensable, 156
  economic consequences in rejection of, 161–63
  economic limitations on, 176–80, 247
  electricity produced from, 22, 23, 106
  environmental hypochondria as obstacle to, 224–26
  environmental superiority of, 27
  ETA-MACRO model for, 163–65
  in First Nuclear Era, 220, 221, 222, 226
  government position on, 19, 25, 57–59
  Great Bandwagon Market for, 20
  increased dependence on, 115, 138
  in international relations issues, 217, 232–33
  in long-term energy security, 155
  map of reactors for, 12–13
  as obsolete, 172–73
  oil supplies vs., 25, 138, 140, 156–57
  other energy sources vs., 138–40, 153–54, 161–63
  as political issue, 217–19, 220–21, 250
  scenario for phase-out of, 163–70
  in Second Nuclear Era, 220, 221, 226
  as symptomatic of larger problems, 69
  U.S. projections for, 176–77
nuclear power plants:
  average capacity factor for, 145
  costs of, 135, 137, 141–42, 144–45, 148
  decline in orders for, 135–36, 142, 145
  decommissioning of, 137
  delays in construction of, 142–43, 145, 152
  designs for, 106
  economic constrictions on, 24, 135
  economic inefficiency of, 137

  financial backing for, 145–46, 150–51
  licensing of, 137, 138, 144, 146, 154, 157
  moratorium on building of, 160
  radioactive gases vented from, 28, 42, 48
  shutdown of, 159–60
  waste disposal problem in, 24, 111–12
  see also specific plants
Nuclear Regulatory Commission (NRC):
  bias in, 104–5
  in Browns Ferry accident, 92
  Cohen's paper reviewed by, 78
  "defense in depth" concept in, 98
  Diablo Canyon plant flaws discovered by, 26
  distrust of, 149
  divisional organizations within, 55
  establishment of, 24
  exposure estimates used by, 71
  health studies needed in, 55–56
  industry complaints against, 136
  in licensing issues, 144, 146, 154, 157
  meltdown estimates by, 76
  nuclear waste proposals by, 126, 130, 132
  public participation encouraged by, 147
  regulatory zeal of, 249
  safety regulations under, 25
  on safety-related incidents, 83
  structural changes recommended for, 96
  in Three Mile Island accident, 25, 65, 84, 144, 147, 226
Nuclear Safety Analysis Center, 96
Nuclear Safety Oversight Committee, 136
Nuclear Science and Society (Cohen), 69
nuclear weapons proliferation, 212–13, 217, 218, 228–33
  commercial nuclear development linked to, 232–33
  early policies on, 228–29
  geopolitical pressures against, 230–31
  nightmare scenarios for, 227–28
  Nonproliferation Treaty controls on, 229–30
  superpower carrots and sticks in, 231
  U.S. policy on, 228–29, 230–31, 232
Nye, Joseph, 227

Oak Ridge National Laboratory, 35, 43, 124, 125, 222

Oconee reactor, 222
Office of Management and Budget, U.S., 138
oil, 150, 160, 171, 184
    conservation of, 186–87
    1973–1974 embargo on, 25, 136, 140
    nuclear energy vs., 138–40, 156–57, 219–20, 223
    reductions in dependency on, 186–87
    solar energy as alternative to, 196, 197
Oklo uranium deposits, 120, 121
Organization of Petroleum Exporting Countries (OPEC), 140, 197
Oyster Creek reactor, 20

Pakistan, 229, 230, 231, 232
Parker, H. M., 35
Pauling, Linus, 32, 61n
Peaceful Uses of Atomic Energy, Fourth Geneva Conference on (1971), 141
Perl, Lewis, 146
Philadelphia Evening Bulletin, 74
plants, fuel from, 196
plasmas, 200–202, 203
plutonium, 17, 20n, 109–10, 113, 116
    in breeder reactors, 175, 207, 208, 209, 210, 212–13, 230
    cancer risk from, 43, 67–68, 256n
    for nuclear weapons, 217
    toxicity of, 77–79
Pohl, Robert O., 123–33
Pollock, Richard, 141–47
potassium-40 (K-40), radioactivity of, 47–48
pressurized-water reactors (PWRs), 22, 45, 87–88, 100, 145
Price-Anderson Act (1957), 19, 85, 102–3, 146
Princeton Plasma Physics Laboratory, 202
Project Sherwood, 204
Public Citizen, Inc., 141
Public Law 96-386, U.S., 200
public opinion:
    consequences vs. probabilities in, 221–22
    disinformation in, 59–60
    environmental hypochondria in, 224–26
    as excluded from industry decisions, 143–44
    industry campaign for favorable attitudes in, 104, 155–56

irresponsible reporting as influence on, 75
naïveté in, 248
on nuclear power, 149, 156, 218, 250

Radford, Edward P., 37
radiation exposure, 27–79
    during airplane travel, 48
    ALARA standard for, 42–43, 45, 50
    at atomic test sites, 27, 34, 76
    of atomic workers, 33, 40–41, 42–43
    benefits of, 52–53, 239
    from brick houses, 48, 70, 239
    as cure-all, 36
    disinformation on, 59–60
    distortions in studies of, 53–55
    early standards for, 35–38
    epidemiological studies on, 28, 50–51
    genetic defects caused by, 28, 29, 32, 38–39, 48, 70, 71, 72–73
    in high-altitude cities, 48, 239
    high levels of, 28, 29–30
    in Hiroshima and Nagasaki, 28, 29, 33, 34, 36–37
    hot particle theory of, 77–78, 256n
    irresponsible reporting on, 73–77, 79
    linear hypothesis of, 29–30, 32, 66, 71–72
    low levels of, 28, 29–30, 43–44
    man-made forms of, 27, 28
    from meltdowns, 102
    from natural sources, 27, 28, 47–48, 70, 225, 239
    occupational guidelines for, 31, 35, 45, 49–50
    other health hazards compared to, 72–73
    quadratic hypothesis of, 30
    quantitative understanding of, 70
    rem measurement of, 28
    scientists' divergent views on, 32–34
    sickness from, 102
    somatic effects of, 38
    standard-setting bodies for, 31, 32, 33, 35–36, 40–41, 50, 56, 71
    standards for, 51, 77
    supralinear hypothesis of, 30, 32, 66
    from uranium mill tailings, 128–29
    in uranium mines, 36
    from waste disposal accidents, 130
    from waste leaks, 74–75

radionuclides, 120–21
radon gas, 36, 43, 48, 128
Rancho Seco reactor, 222
Rasmussen, Norman, 24
Rasmussen report, 24, 100, 222, 225
    Browns Ferry accident unaccounted for in,
        106
    conclusions of, 83, 90
    criticisms of, 84
    federal funding for, 24
    flaws in, 90
    meltdown probability in, 105, 106
    optimism of, 105–6
    probabilistic techniques used in, 89–90
    Three Mile Island accident unaccounted for
        in, 93
reactivity accidents, 21
reactor safety, 81–133
    as "all or nothing" proposition, 98
    containment building in, 88
    coolants for, 87
    debate on, 81–86
    design considerations in, 104
    fuel rods in, 87
    of pressurized-water reactors, 87–88
    primary system in, 87–88
    scram system in, 87
Reagan administration, 175, 200
    breeder reactor program supported by, 215
    deregulation under, 136
    nuclear policy changes under, 137–38,
        148, 154
    proliferation issue in, 232
    solar programs cut by, 193, 194n
    in waste disposal issues, 109, 110, 138
rebuilding, as reactor decommissioning
    method, 137
Reich, Eunice T., 150
Reister, D., 223
Resources for the Future, 162–63
Reynolds Aluminum plant accident, 100
Richels, Richard G., 160–70
Rickover, Hyman G., 19, 21
Rocky Flats, Colorado, plutonium waste at,
    43, 79, 237
Rogovin report, 84, 100
Roosevelt administration, 16
Rose, David J., 174
Rossin, A. David, 152, 155

Roswell Park Memorial Research Institute,
    39
Rotblat, Joseph, 37
Rotty, Ralph M., 220
Russell, Bertrand, 237

Salazar y Frias, Alonzo, 226
SALT negotiations of 1979, 230
Sanders, Barkev S., 53, 54
Sandia Laboratory, 82, 127, 204, 212
Sant, Roger, 182–83
Saudi Arabia, 196, 231
Saudi Arabian Monetary Agency, 135
Sawhill, John, 183
Schlesinger, James R., 147
Schweitzer, Albert, 32
scramming, 22, 87
Seaborg, Glenn, 56
Seabrook nuclear plant, 141, 142, 237
Seligson, Carl H., 151
Shaw, Milton, 20, 21, 22–23
Shippingport nuclear plant, 19, 20, 21
Shiva laser, 203
Shoreham nuclear reactor, 142
social good, concept of, 58–59, 69
sodium coolant system, 210–12
soft energy technologies, 172–73, 189–92
Solar and Conservation Bank, 193
solar energy, 104, 147, 162, 171, 172,
    173–74, 175, 189, 190, 192–98,
    241–42
    barriers to, 195–96, 197
    commercialization of, 197–98
    energy needs fulfilled by, 194–95
    federal funding for, 192–93
    on-site forms of, 195, 196
    plant fuels as form of, 196
Solar Energy Research Institute, 193, 194n
somatic effects of radiation exposure, 38
sorption, 121
South Africa, 227, 229, 230
South Korea, 136, 230, 231, 232
Soviet Union, 18, 98–99, 133, 180, 204,
    215, 220, 229, 231
Spain, 229
Special Power Excursion Reactor Test
    (SPERT), 21–23, 100
steam explosions, 100

Stern, Ted, 205
Stewart, Alice M., 37–38, 39, 40, 41, 53, 54, 225
Strassman, Fritz, 16
Straume, T., 51
Strauss, Lewis, 18–19, 20, 147
submarine reactors, 19, 21, 104
Sundesert nuclear plant, 240–41
Sweden, 182, 190, 220
synthetic fuels, 25, 189, 192
Szilard, Leo, 16

Taiwan, 136, 230, 231
Tamplin, Arthur, 61–63
Taylor, Vince, 155–60, 180
Teller, Edward, 81
Tennessee Valley Authority (TVA), 152
Terris, Bruce, 242
Three Mile Island nuclear plant accident, 25, 27, 92–97, 136
  cancer death risk from, 51–52, 65, 74
  causes of, 92–93, 94–95
  China Syndrome and, 221
  as "close call," 75–76, 83, 84
  complacency dispelled by, 96
  congressional investigation of, 155
  costs of, 137, 147
  defense-in-depth chain in, 98
  deficiencies revealed by, 249
  diagram of, 94–95
  fission products in, 100
  hydrogen gas explosion in, 100
  industry's reaction to, 104, 149
  Krypton gas in, 27, 226, 239
  meltdown potential in, 83, 84, 221
  NRC investigation of, 25, 65, 84, 144, 147, 226
  nuclear generation decline after, 157
  plant operators in, 93
  pregnant women evacuated after, 225
  President's Commission on, 93, 96, 146
  probability of, 222
  protests after, 141
  psychological stress in, 93
  public concern demonstrated by, 103
  radiation exposure levels in, 72, 74
  Rasmussen report refuted by, 105–6
  reactor sales cancelled after, 142

safety mechanisms in, 88, 93, 96
safety steps implemented after, 105, 146
studies on effects of, 33
Tillerson, J. R., 127
Tokamak Fusion Test Reactor (TFTR), 202, 205
tokamaks, 202, 204
Totter, John, 225, 226
tritium, 199, 200, 203, 205, 206
turnkey reactors, 20, 146

Union of Concerned Scientists, 76
United Nations Scientific Committee on the Effects of Atomic Radiation (UNSCEAR), 31, 63, 65, 71, 255*n*
Unsafe at Any Speed (Nader), 141
Ural Mountains, waste disposal accident in, 98–99, 133
uranium, 16–17, 18
  in atomic bombs, 16, 17
  breeder reactor conversion of waste from, 175, 207–8
  in commercial reactors, 20, 22, 207, 214
  depletion of reserves of, 171, 206–7, 213–14, 221
  enriched, 17, 22
  milling process for, 128
  mill tailings from, 128–30, 261*n*, 262*n*
  miners, 36, 43, 50, 67, 129
  in nuclear fuel cycle, 111
  from reprocessed spent fuel, 116
  U-235, 16, 22, 207, 208, 210
  U-238, 16, 175, 207, 208, 209, 210
utility companies:
  coal as oil substitute in, 157
  construction delays in, 142–43, 145, 152
  credit position of, 150–51
  decline in reactor construction by, 135–36
  decommissioning costs to, 137
  financial outlook for, 151–53
  in fusion research, 204–5
  investor-owned, 151
  investors in, 145–46, 151
  limitations on liability of, 19, 146
  radiation effects downplayed by, 44
  rate increases by, 150
  reactor operation opened to, 18

regulatory climate for, 136, 138, 143, 148, 149–51, 155
reserve margin of, 153, 156–57
safety agencies established by, 96
turnkey reactors purchased by, 20

Velocci, Tony, Jr., 148–55
Vietnam War, 235–36, 237
Voelz, George L., 67
von Hippel, Frank, 84, 97

Wall Street Journal, 155–56
Walske, Carl, 148
Warner, I. M., 126
WASH-1400 report, *see* Rasmussen report
Washington Analysis Corporation (WAC), 141
Washington Public Power Supply System (WPPSS), 135
waste disposal, nuclear, 109–33
    from atomic bomb production, 35, 111
    breeder reactors as method of, 175, 207–8
    brine as problem in, 122, 126–27
    for commercial reactors, 24, 109
    containment packages for, 114, 117, 122, 126, 132
    demonstration site for, 114–15, 119, 122–23
    of fission products, 112
    for fusion reactors, 199, 205
    of gases, 42
    glassification method in, 114, 126
    government role in, 111–12, 116
    groundwater problem in, 120–21
    heat problem in, 117, 118–19, 121–22, 127–28
    of high-level waste, 109, 112
    human error problem in, 130–32
    ingestion hazard in, 113, 117
    leaks from, 74–75, 110–11, 120, 121
    local laws against, 24, 110
    long-term hazards in, 112–14, 117, 263n
    of low-level waste, 109, 112, 128, 133
    Lyons project for, 124–27, 130
    from military facilities, 109, 110–11
    monitoring required in, 132–33
    as political problem, 24, 109–10, 112, 123, 218, 221, 240
    public opposition to sites for, 110, 133, 218, 221
    Reagan administration position on, 109, 110, 138
    repository guidelines for, 119–20
    reprocessing vs., 109–10, 116
    retrieval requirement for, 116
    scenario for accident in, 130–31
    sorption as factor in, 121
    subseabed isolation method of, 116
    technical barriers to, 122–23, 133
    techniques for, 116
    "throwaway cycle" method of, 116
    transmutation method of, 116
    of transuranics, 112, 116
    in underground geologic formations, 113–15, 117–18, 119–23, 125–28, 132
    or uranium mill tailings, 128–30, 261n, 262n
Waste Isolation Pilot Plant (WIPP), 119
Wayne, John, 34
Weart, Wendell, 127
Weinberg, Alvin M., 218, 219–26
Westinghouse, 19, 20, 100, 145, 205
West Valley reprocessing plant, 43, 109, 133
Whitman, M. J., 141
Wilson, Carroll, 21
wind power, 171, 190, 193, 194, 241
Windscale reactor accident, 99
Wolfe, Bertram, 219, 238–45
wood, fuel from, 194, 196
World Oil, 172

x-rays, radiation from:
    British study on, 28, 29, 34, 37, 41, 62
    cancer risk in, 37–38, 39–40, 50
    of chest, 48, 50
    levels of, 48, 70
    standards for levels of, 45

Zion Nuclear Power Plant, 101, 102

The Authors

MICHIO KAKU is an associate professor of nuclear physics at the City College of New York. Dr. Kaku has conducted extensive research in unified field theories and has delivered scientific papers throughout the world. He is director of the Institute for Peace and Safe Technology in New York City, hosts a radio show, "Nuclear Alert," and is a frequent speaker on college campuses about the problems of nuclear power.

JENNIFER TRAINER is a writer and free-lance editor. Her articles have appeared in the *Boston Globe,* the *New Bedford Standard Times,* and *M.I.T. Technology Review.* For the past several years, she has been lecturing and writing on the subject of nuclear energy.